The Story
of
Our Fruits and Vegetables

The Story of Our Fruits and Vegetables

by Dorothy Crispo

illustrations by Frank Aloise

The Story of Our Fruits and Vegetables

Copyright © PathBinder Publishing, LLC, 2006
All Rights Reserved. Printed in the United States of America.

No part of this work may be reproduced or transmitted in any form or by any means. Electronic or mechanical, including photocopying and recording, or by any information storage or retrieval system, without the prior written permission of the copyright owner.

ISBN: 0-9776232-7-0

Written by: Dorothy Crispo
Illustrated by: Frank Aloise
Originally Printed in 1968

Publisher:
PathBinder Publishing, LLC
P.O. Box 302
Earlysville, VA 22936
www.PathBinder.com

Dedicated to
ANIELLO CRISPO
a pioneer in American produce.

ACKNOWLEDGMENTS

The author wishes to express her thanks to her brother-in-law, Frank Crispo, for suggesting the book and providing valuable data.

Grateful appreciation is also extended to Wallace Whittaker for his interest and encouragement.

The author is also indebted and wishes to express her thanks to Janice Devine for writing the Introduction, and providing interesting additions to the text.

Appreciation is also extended to Miss Ethel M. Ryan for reading the manuscript, and offering valuable suggestions.

And the author is particularly grateful to Paul Crispo, her husband, for his help throughout the extensive research.

Contents

Foreword	10
Introduction	11

FRUITS

Apples	17
Apricots	21
Avocados	23
Bananas	25
Blackberries	28
Blueberries	30
Cherries	32
Coconuts	34
Cranberries	36
Currants	38
Dates	40
Figs	42
Gooseberries	44
Grapefruit	45
Grapes	47
Lemons	49
Limes	51
Mangoes	53
Melons	54
Nectarines	56
Olives	57

Oranges	59
Papayas	62
Peaches	64
Pears	67
Persimmons	68
Pineapples	70
Plums	71
Pomegranates	73
Quince	74
Raspberries	75
Strawberries	77
Tangerines	79
Watermelons	81

VEGETABLES

Artichokes	84
Asparagus	86
Beets	87
Broccoli	89
Brussels Sprouts	91
Cabbage	92
Carrots	94
Cauliflower	95
Celery	96
Chickory	98
Collards	99
Corn (Maize)	100
Cowpeas	103
Cucumbers	104
Eggplant	105
Endive-Escarole	106
Garden Peppers	107
Garlic	109

Garden Peas	111
Green or Wax Snap Beans	113
Hominy	114
Kale	115
Kohlrabi	116
Leeks and Chives	117
Lentils	118
Lettuce	119
Lima Beans	121
Manioc	122
Mushrooms	123
Indian Mustard	126
Okra	127
Onions	129
Palmito	131
Parsley	132
Parsnips	133
Potatoes	134
Pumpkins	137
Radishes	139
Rhubarb	141
Rutabagas	142
Shallots	143
Soybeans	144
Spinach	145
Squash	146
Sweet Potatoes	148
Swiss Chard	150
Tomatoes	151
Turnips	153
Watercress	154
Yams	155
Bibliography	156
Index	157

Foreword

The purpose of this book is to tell briefly and simply, the origin and history of our American fruits and vegetables. Some of them are natives, but most of them traveled for hundreds of years and over thousands of miles before settling here.

This is the story of their journey, their adventures along the way, and some of the curious customs associated with them.

It is the author's hope that this knowledge will not only be useful, but entertaining, for readers of all ages who perhaps have never thought of fruits and vegetables but as part of their daily fare.

Introduction

The story of our fruits and vegetables is the story of people ... explorers, colonizers, missionaries, conquerors, plant hunters, botanists, frontiersmen ... all the people, known and nameless, who have carried seeds and shoots and plants from the ends of the earth and produced new breeds and varieties in new places.

"As American as apple pie" is a familiar cliché. Yet the apple is not a native American at all. A typical American breakfast starts with orange juice. But oranges came here as foreigners, world travelers whose ancestors grew in the Far East thousands of years ago.

How about that summer picnic veteran, the watermelon? No again! Peaches, pears, parsnips, peas? All immigrants.

Aren't there any truly American contributions to world agriculture? Yes, a number of fruits and vegetables qualify as native, and two of them are so important that they balance everything we have acquired from other continents: corn and potatoes. Large areas of Europe depend on potatoes as a mainstay, while corn is important to many economies.

The agriculture of the United States has undergone dramatic changes within the past fifty years. There is no longer a typical American farm. Perhaps there never was, but a century ago we cherished an idyllic picture of the cozy fields and pastures, the barnyard, all geared to the work of one man and his family and the plodding pace of the horse. Now the fence lines are vanishing as cities and suburbs engulf farmlands and super highways cut through woodlots and pasturelands.

Agriculture has become big business and single crops of fruits and vegetables often stretch for miles. New technologies increase the yields and skilled plant breeders increase size and quality.

Perhaps the most spectacular developments have taken place, and are still taking place, in the American West. There the desert is indeed blooming and once arid areas such as the "Great American Desert" are green with lettuce, carrots, tropical fruits . . . a nearly endless variety to fill our supermarkets and tempt our affluent shoppers.

Yet, the irrigation which makes this possible is not a new miracle.

"The romance of water and dry land goes back long before Columbus came to this country, possibly as early as A.D. 500," say Ladd Haystead and Gilbert C. Fite in their story of the agricultural regions of the United States. "The best account we have centers around the several-stories ruin of a building, probably a combined granary and fort, near the town of Florence, Arizona, in the valley of the Gila River. The name of this building and its surrounding excavated ruins is Casa Grande.

"Here, a thousand years ago, lived a people called the Hohokam, or Water People. They were an agricultural people, not a nomadic pastoral folk, as were most of their neighbors. From what source we do not know they had gained a rather good, albeit primitive knowledge of engineering principles and hydrology. They harnessed the Gila, dug ditches and laterals, and created a garden in the desert. Corn, beans, and melons flourished under their husbandry and the magic of water on soils that were in a condition of unleashed richness, soils that contained the best plant-food values, washed or blown down and deposited in the broad valley."

No one knows what happened to these mysterious Water People, but when it came time to launch the great Salt River Project in Arizona, modern engineers could not improve upon the ancients, whose ditches were so correctly planned that forty miles of Hohokan ditch-line are in use today!

This book is meant to trace the long life stories of the fruits and vegetables we grow within the United States today, where they came from, and, whenever possible, who were the dedicated men and women who developed them into the superior products picked and harvested across our continent in this computerized, jet-propelled age in which we live.

JANICE DEVINE

Apples

The winner of what was probably the world's first beauty contest was awarded an apple! Paris, arch-villain of Greek mythology, had the touchy assignment of choosing the fairest among three goddesses, Hera, Athena, and Aphrodite. Aphrodite won, but, sad to say, the contest was rigged and the apple was blamed as the cause of the Trojan War. All three goddesses tried to bribe Paris, but it was Aphrodite who offered Paris his heart's desire – Helen, most beautiful of women. She helped in the celebrated abduction of Helen to Troy with King Menelaus of Sparta and his warriors in pursuit. Aphrodite's prize became known as the "apple of discord."

The life story of the apple is lost in antiquity. It appears in poetry, legend, and proverbs, all the way from the plea to "Comfort me with apples" in the *Song of Solomon* to "the apple of his eye" and "an apple a day keeps the doctor away."

Like so many of the fruits we think of as typically American, apples had to travel for thousands of years to get here. Scientists believe that the first apples grew in the region stretching from the eastern Mediterranean to the Caspian Sea, including parts of southeastern Europe and southwestern Asia. Stone Age man is known to have developed methods of sun-drying and freezing apples. He also practiced what is now called "vegetative propagation," or the planting of a branch rather than a seed in order to carry on the production of a tree with superior fruit. Four centuries before Christ, several varieties of apples were recorded by the Greek "Father of Botany," Theophrastus, while Cato, Roman botanist, listed seven varieties a century later. Before the year A.D. 100 Pliny had increased the number to 36 different types.

Apple shoots and seedlings came with every shipload of pioneers to the New World, for by then apples had become Europe's single most important fruit. It was not long before apple butter was bubbling in iron cauldrons over wood fires and apple cider was the most popular drink in the colonies. Apples went into jellies, pies, and dumplings and Louisa May Alcott's house in Concord, Massachusetts, was named for the family's favorite dessert, "Apple Slump." Connecticut in its early days was described by an anonymous versifier as "a land of notions, of apple sauce and greens ... a land of pumpkin pies, a land of pork and beans." And when Boston became too crowded for William Blackstone, he collected his prized apple seedlings and in the year 1635 took off for Rhode Island where he built a house near what is now Providence, established a lush apple orchard, and took to riding a white bull around the countryside, giving away apples to any child he met. The fruit of his "sweeting apple trees" was the ancestor of the crisp, snappy Rhode Island apple which still goes into Blackstone Pudding, mixed with white corn meal and molasses, slow baked, and served with hard sauce.

Today apples are grown in every one of our states and in virtually all parts of the temperate world. Apples can survive severe cold, and they can also thrive in a warm climate as long as they have a moderate winter in which to rest.

Many of the fragrant green orchards of the Middle West began with the wanderings of an appealing frontier character called Johnny Appleseed who walked out of the wilderness near what is now the village of Licking Creek, Ohio, in 1801. Wearing an iron pot on his head, he led a horse loaded with burlap bags. Inside the bags were apple seedlings he brought all the way from Massachusetts.

His real name was John Chapman and he described himself as "by occupation a gatherer and planter of apple seeds." For fifty years he floated down the Ohio River and roamed the countryside, everywhere leaving in carefully selected clearings his gift of apple trees. He loved all of nature's living things and it is said he even put out campfires so that the mosquitoes wouldn't get singed!

There are literally thousands of varieties of apples today, yet when the Gold Rush brought hordes of 49'ers to California, that state was so lacking in fresh fruit that a single box of apples shipped down from a pioneer orchard in Oregon was sold for $500! An enterprising Californian auctioned off a small plot of land bearing four apple trees for $800.

Apples frequently are named for the pomologist (or one who practices the science of apple cultivation) who found a tree bearing superior fruit. John MacIntosh, clearing woodland in Dunclas County, Ontario, found the tree which was the ancestor of his namesake apples. In Wilmington, Massachusetts, a stone apple tops a pillar marking the "birth" of the first Baldwin apple. The Winesap dates back two centuries and the Yellow Newtown is also a colonial product, while the Golden Delicious is only a half century old.

As apples crossed the continent in covered wagons or rounded the Horn in sailing ships, it became evident that the fertile Northwest offered the best of all habitats. The state of Washington, especially the Yakima, Wenatchee and Okanogan areas east of the Cascades, produces more apples than any other place in the world, more than 30 million bushels at recent count. However, the old orchards which dotted North Carolina to New England still produce prized apples and still hold gala Apple Blossom Festivals, with pretty American girls crowned Apple Blossom Queens. And people still say "as American as apple pie."

Crab apples are a small, usually very acid variety of the fruit. They are used principally for making pickles and jelly. Some kinds are also grown for making cider. Still others are cultivated as flowering shrubs.

Apricots

The delicate pink blossoms of the apricot first perfumed the springtime air of China some 3,000 years ago. Plant experts seeking the origin of this golden fruit have found a character in Chinese writings, dated earlier than 2000 B.C., which they believe represents the apricot.

Because the apricot blooms earliest of all fruit trees, it was given the Latin name, *"praecoquum,"* meaning "early ripe." Pliny referred to the apricot in his botanical writings, claiming that it was known in Italy around 100 B.C. Before the theory was advanced of its Chinese origin, the apricot was thought to be native to Armenia and was called "prunes armeniaca."

Explorers and conquerors from the earliest times took fruits and vegetables with them, and it is believed that the apricot reached Greece with Alexander the Great in the 4th century B.C.

The earliest known apricots in this country were reported growing in Virginia by Captain John Smith in 1629.

Later they reached California with the Spanish missionaries. California is today the major state where apricots are grown commercially, to be shipped across the country fresh, or to be dried, canned, or frozen or squeezed into nectar.

Apricots belong to the group of drupes, or fruit with hard, woody pits. This is one of nature's ingenious devices for propagation — to surround the seed with an appetizing pulp so that man or beast or bird will eat the fruit and drop the seed. Modern apricot propagation, however, is by graft, and the fruit is plumper, juicier and larger than its colonial prototype.

Avocados — Alligator Pears

Montezuma, ruler of the ancient Aztec Indians of Mexico, served avocados to Cortez and his Spanish conquistadores at a royal reception he gave for them on their arrival at the capital of Tenochtitlán, now Mexico City, in November, 1519.

Montezuma was dead in less than a year, whether at the hands of the conquistadores or of his own rebellious subjects historians are not sure. But the festive fruit he served his Spanish visitors has survived to be a favorite in the United States. Until the 1930's the avocado was an exotic outside the regions of California and Florida where avocado groves were developing slowly. It was not found, even on the menus of posh restaurants such as Delmonico's in New York, in the early 1900's, and it was considered a gourmet delicacy until after World War II, when widely traveled GI's and spreading supermarkets catapulted it into increasing acceptance.

The avocado, or alligator pear, is native to Mexico, Central America, and parts of South America. It was cultivated by the ancient Mayas, Toltecs, Aztecs, Incas, and other great Indian civilizations for centuries before the advent of the white man to the New World. A favorite lunch in Mexico today teams tortillas with mashed avocado, a touch of lime juice added, and strong black coffee.

The early Spanish spelling of the fruit, derived from the Aztec picture sign, was "ahuacatl." Our word "avocado" comes from later modifications.

The Mayan Indians of Guatemala called it "on" and the Incas of Peru, "palta," which is still a popular name in many parts of South America.

Early explorers were enthusiastic about the new-found fruit, and took it back with them to Europe. It was gradually propagated throughout the areas with climates favorable to its culture, and by 1825 was growing in the Hawaiian Islands, Africa, and Polynesia.

The first definite record of the introduction of avocados into the United States was in 1833 when the well-known horticulturist, Henry Perrine, brought some in from Mexico and planted them south of Miami, Florida.

The avocado is very rich in protein and oil. In many parts of the world it is an important food, and a valuable meat substitute. In the United States, however, it is served raw mainly in salads and desserts.

Bananas

Ever wonder which fruit was said to be the "source of knowledge of good and evil" in the Garden of Eden?

According to Roman legend it was the plantain, or "cooking" banana, "musa Paradisiaca," meaning "fruit of Paradise." And the botanical name of the common banana, "musa sapietum," means "fruit of the Wise Men," from a legend that the sages of India reposed in its shade and ate its fruit. Our name "banana" originated much later in Africa.

The banana is one of the oldest fruits known to mankind, and was among the first to be cultivated.

It is believed to be native to southeastern Asia, and it was named in the ancient Chinese, Sanskrit, and Malay languages.

The armies of Alexander the Great found bananas growing in the Valley of the Indus around 327 B.C.

Arab slave traders are credited with carrying the fruit from India across Africa to the coast of the Atlantic. From there Portuguese explorers took it to the Canary Islands.

Bananas were also spread to the tropical Pacific Islands in prehistoric times.

They were introduced into the New World by a missionary priest, Friar Tomás de Berlanga, who brought some plants from the Canary Islands to the Island of Hispaniola in 1516. They thrived well, and spread rapidly throughout the American tropics.

Bananas became big business in the United States because of a ship captain's bad health! Captain Lorenzo Dow Baker, of Wellfleet, Cape Cod, sailed his schooner *Telegraph* toward the warm water of the Caribbean instead of the chilly Grand Banks, explaining that his health was endangered by the winter cold. In Jamaica he took a fancy to the bananas he found in the native market. He tried to take some stalks home with him, but the fruit rotted long before it reached New England. By experiment, he found that green bananas kept in a dark corner of the ship's hold would survive the voyage. Soon he began marketing his "eighteen-day bananas" and finally he stayed in Jamaica while other captains carried the fruit northward to ports such as New Orleans and Mobile. Before the turn of the century he had founded the United Fruit Company. In 1905 the New Englander with the happy allergy to cold weather was honored by the Jamaican government which said:

"In thirty years Captain Baker has done more for Jamaica than the British Empire in three hundred years."

Bananas do not grow on trees. The plant is a gigantic herb. When fully developed it has a palm-like appearance. Its large leaves protect it with an umbrella-like action. When the sun is intense, they fall around the plant to keep it from too great evaporation. In wet or cool weather they curve up to permit increased evaporation necessary for the rapid growth of the plant.

A banana stalk is made up of clusters, which are called

"hands." Each "hand" has from 10 to 20 individual bananas, called "fingers."

The banana is one of our most important tropical fruits. Not only is it very flavorful, but it possesses great nutritional value. Together with the plantain, or "cooking" banana, it feeds hundreds of thousands of people throughout the world.

Many jungle and tropical animals and birds live on bananas. Even the mules used for carrying the fruit from plantation to railroad sidings for shipment enjoy them so much that often nose baskets are necessary to prevent them from nipping off a tasty "hand."

Blackberries

The blackberry has been enjoyed by man from earliest times. Like the raspberry, it is a bramble bush and belongs to the rose family.

It is native to the temperate regions of Asia, North America, and Europe.

While there are many species of blackberries, they may be divided into two general classes: those that grow on erect upright plants, and those that grow on trailing or prostrate plants. These latter are also called dewberries, and are the more delicately flavored of the two.

European blackberry culture is fairly new, dating back to the beginning of the 20th century. In the Americas it was started somewhat sooner. This was undoubtedly due to the fact that blackberries were so abundant in their wild state that mention in early agricultural papers was mainly advice on how to kill the plants to clear the ground for farming.

Some of our leading varieties of the trailing type are the Youngberry bred by B. M. Young of Louisiana; the

Loganberry, which originated in the garden of Judge K. H. Logan of Santa Cruz, California, apparently as a cross with the raspberry; and the Boysenberry, of chance origin in California, and which is similar to the Youngberry.

Blackberries have also been cultivated for many years in South America, and some, such as the Colombian kind, are as long as 2 inches and 1½ inches thick.

In addition to their many uses raw, blackberries are delicious in jams, jellies, pastries, ice creams, and sherbets. They are also used for making wines and cordials.

Blueberries

From tropical tribes to the Eskimos, and practically all areas in between, the blueberry has for thousands of years been a favorite food of man. It is perhaps the most widely distributed of all fruits.

Because of its abundance wild, breeding and cultivation of blueberries are relatively new.

The name "huckleberry" is often used interchangeably with that of blueberry. Huckleberries, however, although belonging to the same family group, are smaller and have large bony seeds. Blueberries are sweeter, and their seeds are so small that they are barely noticed when eaten. Of the two types, only blueberries are cultivated.

The huckleberry is part of our heritage of American Indian fruits. At the annual feast of Cherokee foods held at the museum in Cherokee, North Carolina, huckleberries share the honors with fox grapes, ground cherries, and some thirty other dishes. Cherokee women are said to have encased huckleberries in a corn-meal dough and deep-fried them in bear fat.

A New Jersey woman is partially responsible for the plump, nearly seedless blueberries which come to us fresh, canned, in pies and muffins. Every year Elizabeth C. White, of Whitesbog, New Jersey, offered prizes for the high-bush plants bearing the largest berries. Hearing of her work, Dr. Frederick V. Coville, U. S. Department of Agriculture botanist, began in 1909 to cross superior berries he had found. The result of their pioneer work is the 18 varieties we find on the market today.

Whether specially bred, or gathered wild, both berries are great American favorites.

Cherries

Miniature hatchets and candy cherries perpetuate Parson Weems' tall tale that young George Washington chopped down his father's favorite cherry tree, and then bravely confessed to the crime.

Actually, cherries are a far more important part of America's past than the Washington cherry-tree story. They came on the Mayflower and flourished in New England orchards, and they were planted in California's Spanish missions. Frontiersmen carried them west in saddlebags and covered wagons. Henderson Luelling, arriving in Oregon in 1847, planted the ancestors of that state's enormous sweet-cherry industry.

In Traverse City, Michigan, an annual Cherry Festival celebrates blossoming miles of tart red cherries, begun in the 1890's when a newcomer named B. J. Morgan bought a few sandhills and discovered that his cherry seedlings did remarkably well there. Today that section of Michigan produces more than 100 million pounds of the makings of tarts, ice cream, pies, soft drinks, jam, candy, and cordials.

Cherries have been cultivated and enjoyed by man since earliest times. Cherry pits have been found in the Stone Age Swiss lake dwellings, in ancient Scandinavian deposits, and in cliff caves of prehistoric inhabitants of America.

Botanical historians believe that the cultivation of cherries began in China, whose agriculture dates back over 4,000 years.

Theophrastus described cherry trees about 300 B.C. in his Greek botanical records.

Pliny, Roman naturalist of the 1st century, wrote of several kinds of cherries, and a book on farming, written about 50 B.C. by the Roman author Marcus Terentius Varro, includes cherry culture.

The cherry is divided into two general groups; sweet cherries and sour cherries.

Sweet cherries are enjoyed for table use, for canning, and for making maraschino cherries. Two of the most popular varieties are the juicy yellow "Royal Ann" and the large, luscious, dark "Bing." Sour cherries are used primarily in pastries, and for canning.

Still others, such as the beautiful Japanese flowering kind featured in our Capital, have been cultivated for centuries in the Orient for ornamental purposes.

Coconuts

Coconuts sailed the high seas long before Columbus. The large pod holding the nut is buoyant and waterproof ... a miniature ship ... and so it traveled throughout the tropics, landing on small islands and low shores to become one of the world's most useful plants.

The coconut is believed to have originated in southern Asia and the Malay Archipelago. Some authorities say that it is also native to the American tropics.

It has been cultivated for over 3,000 years, and was propagated throughout tropical regions during earliest times.

Today, in addition to being a most valuable economic product of the tropics, coconuts are an important staple article of food and beverage for millions of people.

The wood of the tree is used to build houses and to make furniture. The leaves, or fronds, are used to thatch dwellings, make fences, fashion oars, etc. Coconut oil is expressed from the dry nut; medicines are extracted from the tissue of the fruit, and the shells are made into plates, bowls, and spoons. The dried fronds are used for fuel, and torches at night.

The tree, towering 60 to 100 feet, is crowned with frond-like leaves. It matures in approximately seven years. A carefully tended tree may yield as many as 200 coconuts a year, and produces nuts for about 75 years.

When the nut is green, the flesh inside is soft, and may be eaten with a spoon. The cavity inside contains the coconut milk, which is a refreshing beverage. As the nut turns brown, the meat becomes hard, and the milk dries up.

Gathering or "tumbling" coconuts is a very specialized art. Tumblers are trained from childhood to climb the tall trunks of the palm and cut down the clusters of fruit with their machetes.

Coconuts are exported in large quantities, and we enjoy them in many of our favorite desserts and candies.

Have you ever held a coconut and wondered how to open it? Just look for the "eyes," which are three soft spots on the top of the shell. Pierce these and drain out the milk. Then with a hammer tap the hard shell all over until it cracks and falls off. Or you can break off this shell by heating the coconut in a 350-degree oven for about 30 minutes.

Cranberries

In early days, sailors of North America took casks of fresh cranberries with them on their long voyages at sea. They ate them to prevent scurvy, in much the same manner as sailors of England and southern Europe used limes and lemons.

Cranberries are native to the swampy regions of the temperate and arctic zones of Europe and North America. The first white settlers to our eastern North American mainland found them growing abundantly.

The Indians valued cranberries highly, and used them both as food and as a poultice for blood poisoning. They called them "I-bimi," "bitter berry."

Our name "cranberry" is believed to have originated with early colonists who called the fruit "crane berries," because it was a favorite of the cranes.

For more than 200 years cranberries were harvested from the wild vines, and it wasn't until early in the 19th century that cultivation of the fruit was begun.

Cranberries were used as "hush money," or more properly, perhaps, as a diplomatic ploy in 1677 when ten bushels were sent by Massachusetts colonists to Charles II, along with 3,000 hogsheads of hominy and 3,000 codfish. The King had sent an angry protest when he heard that the fledgling colony was coining its own money, "pine-tree shillings," without royal permission. The shipment of "the choicest products of the colony" was meant to appease His Highness, and apparently it did.

Many years later the potential of cultivating cranberries

was recognized. Henry Hall, of Dennis, Cape Cod, happened to clear a patch of swamp when a sand slide buried his wild cranberries. To his surprise, the vines responded with a vigorous growth. A newspaper story of 1852 reported that he was producing 70 bushels per acre, and the idea of sanding cranberry bogs was soon adopted by Henry's neighbors. Cranberries are still a prize crop on the Cape.

About the same time Henry Hall was experimenting with the cultivated cranberry, John Webb, a New Jersey schoolteacher with a peg leg, drained a swamp and raised the state's first commercial crop, which in 1845 went aboard a fleet of whaling ships.

The test of a good cranberry is said to be in its bounce. In early times, the berries were rolled down a series of steps. The good ones, being firm, bounced like tiny rubber balls; the damaged ones, because they were soft, stayed on the steps. The grading machines used today for sorting the fruit are based on the same principle, and in our completely mechanized era it is still the old wooden cranberry scoop that does the job of "combing" the berries from the vines.

We are all inclined to associate cranberries only with our holiday turkey. Actually, they are delicious all year round with almost every type of meat. And cranberry juice is a crisp, tasty way to get your needed vitamins.

A smaller-sized cranberry grows abundantly in the Scandinavian countries, where it is highly esteemed.

Currants

Puddings should be
Full of currants, for me:
Boiled in a pail,
Tied in the tail
Of an old bleached shirt:
So hot that they hurt.

So wrote versifier Richard Hughes in the early 1600's, and for over 300 years many New Englanders have agreed with him. They brought the tart red currant from England to the Massachusetts Bay Colony in 1639, made it into boiled puddings and jellies to go with venison.

Currants belong to the gooseberry family, and have a similar background and history. Both are native to the colder parts of Europe, Asia, and America, and have been enjoyed by man in their wild form from earliest times.

German writers described currants late in the 15th century, and there are references to them in English writings around the middle of the 16th century.

Black currants are not common in the United States, but are grown extensively in Europe. They are said to have medicinal value, and are noted for their high Vitamin C content.

Dried currants are really not currants at all, but a fruit known as the Corinth grape. This is a small, seedless grape which got its name from Corinth, Greece, where it was first cultivated. It has long been grown for drying purposes.

Unfortunately for both currant and gooseberry enthusiasts, these plants are said to harbor a fungus that attacks one of the most valuable trees in this country: the white pine. The planting or shipping of these berries, therefore, is prohibited by law in many of our areas.

Currants are especially popular in jams, jellies, and pastries.

Dates

So important is the date to the Arab world that on the Persian Gulf island of Bahrein all the date palms belong to the Shaikh and his family. It is, in fact, a blessed tree. Mohammed said of it:

"There is among the trees one that is pre-eminently blessed, as is the Moslem among men; it is the palm."

It is one of the world's oldest plants, but as yet its origins are unknown. In the barren desert lands where it thrives, the date palm well merits Mohammed's accolade. It shades the oases, its leaves are used for baskets, bags, ropes, fiber, mats, and it has often been called "the candy that grows on trees" because the date contains half its weight in sugar. Trees nearing the end of their long lives — and they may bear fruit for two centuries — are tapped for a sap which is made into a beverage which has been called "the drink of life." Finally, the ancient trunks are used for fuel. Roasted date seeds are sometimes used as a coffee substitute.

There were date palms growing around the early Spanish missions in California, but the real start of date culture in the United States was in 1890, when our Department of Agriculture imported several varieties of palms from Egypt and tried them out in desert valleys of Arizona and California. More than 5,000 acres of date palms are now grown, most of them in California's Coachella Valley, but dates are so much in demand that quantities are still imported from the Arab world, and we, in turn, send them our latest technologies of date culture.

Travelers in the Coachella Valley area have been known to ask what is inside the "packages" wrapped in heavy paper, dangling from palm trees, and why helicopters hover low over the groves. The paper guards the bunches of dates, many bearing 1,000 or more individual fruits, from insects and moisture, while the role of the helicopters is to fan away dampness.

DATE TREE

Figs

In many southern European countries figs are thrown at newlyweds for happiness and good luck, in much the same manner as rice is in the United States.

Figs are native to western Asia and the Mediterranean area, and are one of the earliest fruits known to man. They are mentioned in the Bible with the story of the Garden of Eden, and fossil remains of figlike plants dating back long before the Stone Age have been found in France and Italy.

Buddhism was born under a fig tree. Actually, the tree was called Bo, but it is a species of fig and under it some time between 600 and 500 B.C., in what is now a jungle region on the border of Nepal, there rested a man called Siddhartha Gautama. The revelations that came to him on that far-off day were the foundation of the great religion, Buddhism, which emphasizes the same quality of quiet meditation Gautama found under his fig tree.

The Romans considered the fig a gift of the god Bacchus, and regarded it as sacred. So did the ancient Greeks and Egyptians, and there is a drawing of a fig tree on the wall of a 12th dynasty Egyptian grave (circa 1989–1776 B.C.).

Figs were known and highly esteemed in Crete in 1500 B.C., and in Greece a little later. Aristotle, the Greek philosopher, wrote of figs in the 4th century B.C.

The Spaniards brought figs to the Americas, planting them first on the Island of Hispaniola in 1520. By the end of the 16th century they were growing abundantly in many parts of South America, and also at St. Augustine, Florida. In 1629 that tireless reporter, Captain John Smith, found a "Mistress Pearce" in Jamestown, Virginia, who had harvested "more than a hundred bushels of excellent figges." The missionary fathers took figs to California and planted them at San Diego Mission in 1769. The California "Mission Fig" derives its name from this fact, and today it is still the leading black fig in that state. In 1839, "new figs" were on the menu of New York's elegant Astor House.

There are many varieties of figs, but they may be divided into two general classes: the light colored figs and the black ones.

It took an immigrant from abroad to make commercial growing of Smyrna figs profitable in the United States. The immigrant was a tiny wasp, the *Blastophaga*, which lives in the fruit of the caprifig. Our vigilant Department of Agriculture experts discovered that the caprifig had long been used in Europe and Asia for pollinating the Smyrna fig. The wasps breed in caprifigs. The fruit with the wasps inside is hung in paper bags in the trees bearing Smyrna figs. As the female wasp crawls from the fruit, she is dusted with pollen. She lights on the Smyrna fig blossoms to lay her eggs, thus pollinating the flowers.

Figs are enjoyed fresh in the localities where they are grown, and in some Mediterranean countries are a basic part of the diet.

They are difficult to ship, however, because they are fragile, and most of the commercial fig production is marketed canned, preserved, or in dried form.

Gooseberries

Gooseberries are native to the colder sections of Asia, Europe, and the Americas. They were cultivated in England and Europe from the beginning of the 16th century. The English were particularly enthusiastic about the fruit, and had yearly gooseberry shows to encourage their development.

European gooseberries were brought to this country by early colonists. They did not thrive well, however, except in the cool, dry regions of the Pacific Coast. The important varieties grown here today are the results of crossing of the European and the native American kind.

Gooseberry shrubs are said to harbor a fungus that kills the valuable white pine trees. This fungus came into the United States between 1898 and 1910. For this reason, the cultivation of gooseberries is prohibited in many sections of the United States.

The name "gooseberry" is believed to have been given the fruit because it was customary to serve it with goose. Others state, however, that it is a corruption of the Dutch name "kruisbes," which means "cross-berry."

Gooseberries are prized for making pies, puddings, jellies, jams, and pastries.

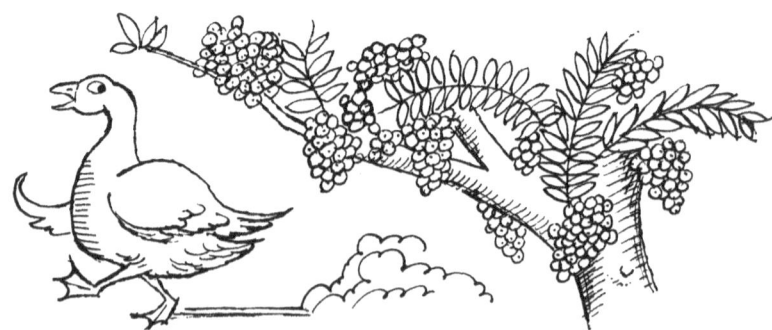

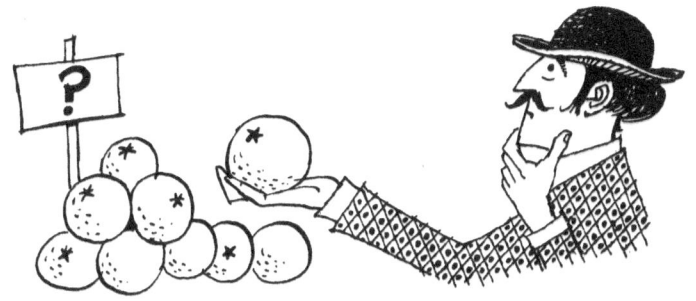

Grapefruit

"Nobody knows what the darned stuff is."

The "darned stuff" was a small shipment of grapefruit being offered to the public in the 1890's by a venturesome fruit dealer. His customers stared and prodded, but nobody bought until a visiting Floridian happened along.

"Well, grapefruit," he said. "How much?"

"Anything you'd like to pay, just so's you take them off my hands," was the answer.

Even in 1897 a New York fruiterer felt that he was taking a chance by bringing north 25 boxes of grapefruit. He was shrewd enough to offer them as gourmet delights and he managed to get $156.25 for the lot.

That is how new the now ubiquitous grapefruit is to the American table. Yet the fruit, in its present juicy, sometimes nearly seedless form, is distinctly a product of the Americas. It was our canny citrus growers who grafted and bred until their groves were producing a superior, delicious fruit. One Florida grower, however, found that he couldn't find a market for the new pink grapefruit he had developed. It was left to a Texan to adopt the pink grapefruit and make it an important crop. Actually, the town of New Mission, Texas, calls itself the "Home of the Grapefruit" and celebrates with a citrus fiesta in late January. The celebration is as much in honor of the pioneers who refused to quit when their crops were ruined time and again by freezes, as it is to the fruit itself. Arizona, thanks to irrigation, is finding its place as a prime grower of grapefruit.

The ancestor of the grapefruit, the pummelo, or shaddock, is believed to have originated in the Malay Archipelago, where it has been cultivated for over 2,000 years.

The pummelo is a large fruit that looks like a very thick-skinned grapefruit. It apparently reached Europe around the middle of the 12th century. It was known there as "Adam's apple," and was grown mainly as a garden curiosity.

Captain Shaddock, of an East Indian ship, is said to have introduced the pummelo to the Americas. He stopped at Barbados, an island in the British West Indies, on his way to England, and while there left some pummelo seeds. The fruit, thus, was given his name.

Shaddocks were recorded in 1696 in a catalogue of plants of Jamaica, another island of the British West Indies.

Grapefruit, as we know it today, originated in the West Indies, probably as a mutation of the shaddock, or a cross between the shaddock and the sweet orange.

The name "grapefruit" was given the fruit in Jamaica, because it was thought to resemble the grape in flavor, and also because it frequently grew in clusters.

A Spanish nobleman, Don Philippe, introduced grapefruit into Florida from the West Indies about 1840. Most of our present varieties come from that original crop.

Grapefruit did not become generally popular until around 1915. Canned grapefruit and canned and frozen juice, almost as important as the orange to the American breakfast, were virtually unknown until the 1930's.

Grapes

North America was once named after the grape. The Norsemen, said to be the first European visitors to this continent, found native grapevines so abundant that they called the country "Vinland."

Only a few varieties of the grapes we know today originated in this country. The scuppernong is one, and for many years the "mother vine," which grew wild on the shores of North Carolina and was transplanted by two elegant Elizabethans, Philip Amadas and Arthur Barlow, to the island of Roanoke, spread its shade and fragrant fruit over more than an acre. Amadas and Barlow went back to England to report that the new land was "the most plentiful, sweete, fruitful, and wholesome of all the world." That was in the summer of 1584, but when Sir Walter Raleigh sent investigators in 1591 to see how the Roanoke Colony was doing, every one of its 118 men, women, and children had vanished, leaving only the scuppernong grapevine as a memento; that, and a name carved on a tree — Croatan — a powerful Indian chief of the region.

In California, the famous Mission of San Diego houses

the nation's oldest winery. The great, gnarled Trinity Vine bore grapes for more than 170 years. California produces excellent wine grapes and thousands of visitors annually tour the vineyards and wine cellars up and down the coast.

Famous grapes such as the delicate Niagara White and Delaware Pink flourish around the Great Lakes. Paw Paw, Michigan, welcomes people to vineyards and grape-pressing plants in late October when the countryside blazes with autumn color.

That many grapes came to us from Europe is attested by the carved wooden mantels in the old colonial houses of Massachusetts, where designs of grapevine and breadfruit predominate.

The delicious blue Concord grape was developed by a visionary who came to Concord, Massachusetts, in the early 1800's "to see what I can find among our wildings." A shoot from his original vine still clambers over the historic Ephraim Bull cottage, and because he never tried to make any money from his Concord, the epitaph on his tombstone in Sleepy Hollow Cemetery says simply:

"He sowed . . . Others reaped."

Grapes have been cultivated from such ancient times that it is not known where they originated although it is believed to have been in the area about the Caspian and Black seas.

Before history was put in writing, grapes were being planted and grown in a manner that indicated a long background of knowledge.

Grape seeds found in Swiss Lake dwellings date back to the prehistoric Bronze Age. They are also mentioned in the Hebrew, Greek, and Roman writings, and the Bible tells of Noah planting a vineyard. "And Noah began to be an husbandman and he planted a vineyard." (*Genesis 9:20.*)

Lemons

Our delicious lemonade, early records tell us, was actually invented by the Mongolians in the year 1299. But pink lemonade originated in New Jersey, according to Billie Griffiths, a circus lemonade vendor, of Three Bridges. Billie contended that one day when he was stirring up his tubful of cold lemonade for the afternoon performance, a pair of pink tights drying on a line outside the lady acrobat's tent blew off and landed damply — and pinkly — in his tub.

Lemons, like other species of citrus fruits, are believed to be native to southeastern Asia, and have been cultivated for thousands of years.

Earliest Chinese references to lemons are in A.D. 1175 when the "li-mung" fruit was described.

The Arabs took lemons into Persia and Palestine, where they were growing at the time of the Crusades. Since the fruit is not mentioned by European writers until after that time, it is assumed that the returning Crusaders in the 12th century carried lemons back to Europe with them.

History tells us that when Columbus made his second voyage to the New World in 1493 to establish a colony, he stopped at the island of Gomera in the Canary Group, and while there obtained seeds of oranges, lemons, and many vegetables. He planted the first lemons in the New World in 1493 in his settlement of Isabela, on the island of Hispaniola (Haiti-Dominican Republic).

Later Spanish conquerors carried seeds from these thriving groves to the mainland of Mexico and Central America, and also planted them at St. Augustine, Florida in 1565. Our first California grove was started by the Franciscan padres when they established the San Diego Mission. Today practically all the lemons grown commercially in this country come from California.

Portuguese colonists took lemons with them to Brazil, from where they were spread south.

We favor the acid type of lemon in the United States, but there is also a sweet variety that is prized in Oriental countries.

An interesting note about lemons is that unlike other fruits, harvesting goes on at all seasons. Lemons are found on the tree in all stages of maturity, blossoms often being seen on the same branch with fruit ready to pick.

Lemons have an endless variety of uses as a flavoring for meats, fish, desserts, and other dishes. They are also the basis of many of our "ade" drinks.

Their richness in vitamin C makes them a valuable addition to our diet, and in fact, for many years they were an important food on ships in the prevention of scurvy, a disease caused by lack of this vitamin.

Limes

British sailors were first called "limeys" because of the quantities of limes carried for them on ships to prevent scurvy, a disease caused by lack of Vitamin C. Limes, today, along with their close "cousins" lemons, are still among our most important sources of Vitamin C.

Like other species of citrus fruits, limes are believed to be native to southeastern Asia, and have been cultivated by man for thousands of years.

The Arabs in remote times called them "limoon" and took them, with lemons, to Persia and Palestine. Crusaders of the 12th century found them growing there, and carried the fruit and seeds back to Europe with them. The fruit was called "lima" as early as the 13th century. Sir Thomas Herbert is attributed with having first used the name "lime." He spoke of finding "oranges, lemons, and limes" on the island of Mohelia off Mozanbique during a voyage in 1626.

When Columbus made his second voyage to the Western Hemisphere in 1493, to establish a colony, he stopped at the Canary Islands and obtained seeds of various vegetables and fruits, including limes, to take with him. He planted the limes on the island of Hispaniola, in his settlement of Isabela.

Later, Spanish conquistadores took limes from these groves with them to the mainland, and planted them at St. Augustine, Florida.

Limes were introduced into California by the Spanish Franciscan padres, who planted them along with other fruits at their missions.

Due to the fact that limes require a hot, tropical climate, the bulk of our present supply comes from Mexico and the West Indies. Egypt, however, leads the nations of the world in lime production, with both the sour kind popular with us and also a sweet variety which is favored in many places.

Limes are used in nearly every part of the world today to flavor foods, and as a basis for popular "ade" drinks.

Mangoes

The mango is one of the most important and delicious of all tropical fruits.

Because it requires a warm, humid climate, it does not grow in the United States, except in parts of Florida. It is, however, being imported in greater quantities each year, and may soon be a familiar treat in all of our markets.

Mangoes originated in Southeast Asia, and have been cultivated in India for over 4,000 years. Not only were they important as food, but they figured prominently in Hindu mythology and religious ceremonies. Buddha is said to have been presented with a mango grove as a place of repose, and Akbar, northern Indian emperor of the 16th century, had over 100,000 mango trees planted in his garden at a time when large orchards were most unusual.

The Portuguese navigators carried mangoes to East Africa, and from there introduced them to the Americas by way of Brazil. Today, they grow in great profusion throughout the tropical countries.

Mangoes grow on a large tree, sometimes 40 feet high, with thick foliage of big shiny leaves. The bright yellow and red fruit hangs like a pendulum from long stems.

There are hundreds of varieties of mangoes. The finer kinds are sweet and luscious, and taste somewhat like a peach. The less desirable, however, tend to be coarse and fibrous.

In order to preserve their delicate flavor, mangoes should not be cut until just before serving. Halve them lengthwise, peel, remove the large stone, and serve. And of course they may be just peeled and eaten out of hand.

MELOPEPO

Melons

Most people think of melons as fruit and cucumbers as vegetables. Yet they both belong to the same genus, *Cucumis*, with similar structure and growth habits.

The term muskmelon applies to all the popular melons in our markets today, with the exception of the watermelon, which is related to the citron. It is native to Persia and adjacent areas and its name means "perfumed, apple-shaped melon." "Musk" is Persian for a kind of perfume, and "melon" is French from the Latin *melopepo*, meaning "apple-shaped melon," and came from Greek words of similar meaning.

It is also said to have grown wild in India and in Egypt, and the oldest record of the muskmelon is an Egyptian picture of 2400 B.C. in which a fruit appears which has been identified by experts as being a muskmelon.

Melons were introduced into China and the Mediterranean areas of Europe at the beginning of the Christian era. Pliny the Elder, Roman naturalist of the 1st century A.D., described them as "a new form of cucumber called Melopepo."

The Greeks were apparently familiar with the muskmelon as early as the 3rd century B.C., although their first recording of it was by Galen, the Greek physician, who wrote of its medicinal qualities during the 2nd century A.D.

Muskmelon seeds were brought to the New World by Columbus on his second voyage, and planted on the island of Hispaniola. The melons were so well liked by both colonists and the Indians that they were soon being grown throughout the American continents.

The Bible mentions melons (*Numbers 11:5*) among the foods of Egypt the Jews, under Moses, yearned for during their journey in the wilderness.

There are many varieties of muskmelon, including the cantaloupe, honeydew, casaba, banana, Santa Claus, Montreal, and Persian types.

The cantaloupe got its name from the castle of Cantalupo, the country seat of a 16th-century pope, where it was developed from a variety of muskmelon brought from Armenia.

The true cantaloupe is grown in Europe. What we term "cantaloupe" in the United States is the netted or nutmeg muskmelon, which originated in Persia. The major cantaloupe-producing region is California's Imperial Valley, where special strains have been developed to resist powdery mildew.

The Persian melon resembles a large cantaloupe. It is named for Persia, the country of origin of muskmelons, and along with the cantaloupe is a great favorite in the United States.

Honeydew melons are still another distinct variety of the muskmelon. It earned its name from its sweet mildly scented fruit, the flesh of which is a pale greenish-yellow. It is the descendant of an old French variety called "White Antibes Winter," and was introduced into this country about 1900.

All of these luscious varieties of the muskmelon are not only delicious, but are excellent sources of vitamins A and C, and also provide an assortment of other valuable nutrients.

Nectarines

"Nekter," or nectar, in Greek and Roman mythology was the drink of the gods. It was described by Homer as being crimson in color, delectable in flavor, and having the power to give perpetual youth. Because of its delicious flavor, the nectarine derives its name from the word "nekter."

The nectarine is a smooth-skinned or fuzzless peach which at one time was thought to be a different fruit. It is one of nature's most interesting mutations; that is, the parent tree may suddenly produce a fruit foreign to itself. Nectarines may appear on peach trees, and peaches may appear on nectarine trees; or the peach tree may bear all peaches and the nectarine tree all nectarines.

Both trees look alike, and have the same kind of leaves and flowers. The fruit of the nectarine, however, is usually smaller, firmer-fleshed, and tastier. It is also harder to grow than the peach, and more perishable to ship.

The history of the nectarine goes back to the Christian era, then merges with that of the peach, which has been known and cultivated by man for thousands of years.

The Spanish colonists took nectarine seeds to South America, the French, to Louisiana, and the English to the eastern North American mainland.

Nectarines may of course be used in any of the ways that peaches are — and a sweet, ripe, cold one is indeed the "nectar of the gods!"

Olives

To Noah, the olive branch carried by the dove meant that calm had come to the troubled waters. Even in those remote Biblical times the olive was a symbolic and important fruit. It gave its name to the Mount of Olives in the Holy Land, site of many dramatic events in the Old and New Testaments, climaxing in the Ascension of Christ.

The city of Athens in Greece was named in honor of the goddess Athena, after her gift of the olive to mankind.

The olive tree, a small evergreen, is native to Asia and the East Indies, but it was first cultivated thousands of years ago in the area of Syria and Greece. Its species is known as *Olea Europaea*, but our common name "olive" derives from the Latin, *oliva*. Romans prized olives not only as food, but as oil for their lamps and to annoint their bodies, and an old Roman prescription for a long, pleasant life was "Wine within, oil without."

It takes an olive tree eight years to bear its first fruit, but the trees have been known to live for 1,000 years. Italy, Spain, and Greece are major sources of olives and olive oil, and it was from Spain that the first olives arrived and were planted in 1769 at the historic mission of San Diego in California. Our mission olives are all descendants of those pioneer trees. Olives are also being grown in a limited quantity in Arizona, but there is such demand for both pickled olives and olive oil that we still import

quantities from abroad. World production of olive oil is estimated at more than a million tons a year.

Green olives are picked early; ripe, they turn a purply-black, but in both stages they are very bitter and must be soaked in lye solution, thoroughly washed, then soaked in strong salt solution. Pickled green olives are often stuffed with pimientos or anchovies. Both green and black olives are favorite appetizers and cocktail snacks.

Oranges

At the turn of the century American children awaited Christmas morning eagerly, because often it meant an orange tucked into the stocking. Yet in the 1760's a quantity of Georgia orange juice was exported to England!

Thus the paradoxical story of what we now regard as an all-American fruit. The Georgia orange juice was the result of a daring experiment in Savannah, Georgia, one of this country's few planned cities. There the town trustees planted ten acres with apples, peaches, oranges, pears, olives, grapes, pomegranates, and even coconut palms. Georgia's dream of diversified agriculture was doomed as cotton growing, slavery, and the plantation system gradually took over. Cotton was quick, easy money. Food crops were largely forgotten.

Oranges are one of our oldest known fruits. Chinese writings that date back to about 2200 B.C. mention them. The book *Ku Yung*, a tribute to the Emperor Ta Yu, says: "The baskets were filled with woven ornamented silks.

Yet it took the orange at least 4,000 years to reach the New World. Columbus on his second voyage stopped at the island of Gomera in the Canaries and took seeds of oranges and lemons with him to Isabela, his colony on the island of Hispaniola (now Haiti and the Dominican Republic) where 30 years later citrus trees were said to be "beyond counting." In the early 1500's oranges journeyed to South America and Mexico. When St. Augustine, Florida, was settled in 1565 the sour orange was among its fruits. The Florida Indians evidently fancied the new fruit, as white settlers in 1764 described wild orange groves 40 miles long. By 1821, when Florida joined the Union, oranges were beginning to be developed commercially.

The ancestor of the orange is now believed to have originated in South China and Indochina, and it was known to be thriving in Burma 40 centuries ago.

Chinese writings that dated back to about 2200 B.C. mention oranges, and their cultivation in China was far advanced by the Middle Ages. They were not introduced into Europe, it appears, until after the beginning of the 15th century.

The word "orange" goes back to the old Arabian "naranj," and Persian "narang," used before A.D. 300. The Bavarian philosopher Albertus Magnus (A.D. 1193–1280) used the term "arangus" for sour oranges, and from this our "orange" was derived.

Spanish missionaries brought oranges to California from Mexico, and planted groves at their many missions.

Scientifically trained orchardists today control the quality and even the sweetness of our oranges, which are primarily the *Citrus cinensis* or sweet orange. But it was a woman's curiosity that brought to California its fabulous navel orange, sweet and juicy, and a favorite for eating out of hand. In 1875 Eliza Tibbetts of Riverside, California, read a small item about a new orange which had been developed in Brazil. She wrote to Washington and our Agriculture Department secured for her the first two navel orange seedlings.

In addition to sweet oranges, there are the mandarins, with thin, loose peel. Our tangerines belong to this group. The sour oranges, probably the first to be brought to the Americas (*Citrus aurantium*), are used principally for marmalade and some ade drinks.

Orange juice, our breakfast bonanza, did not become a truly national American "must" until the cencentrated juice was frozen, and that was in the late 1940's. Canned orange juice was not even marketed until the 1930's.

Southern California and Florida produce most of our oranges, with an increasing harvest coming from Texas and parts of Arizona. This is because the orange cannot thrive in areas where temperatures fall below 20°, and when, as does happen, a freeze descends on orange groves, smudge pots are lighted and anxious growers spend days and nights to save their precious crops.

Oranges arriving in our supermarkets today have been tenderly picked by workers wearing cotton gloves; they have been washed, brushed, and dried in wind tunnels, and they bear little resemblance to the sour, seedy fruit which began its travels from the Orient so long ago. They are such an important part of our economy that San Bernadino, California, stages a ten-day National Orange Show in mid-March, and Winter Haven, Florida, sets aside a week in February to entertain a million festival visitors.

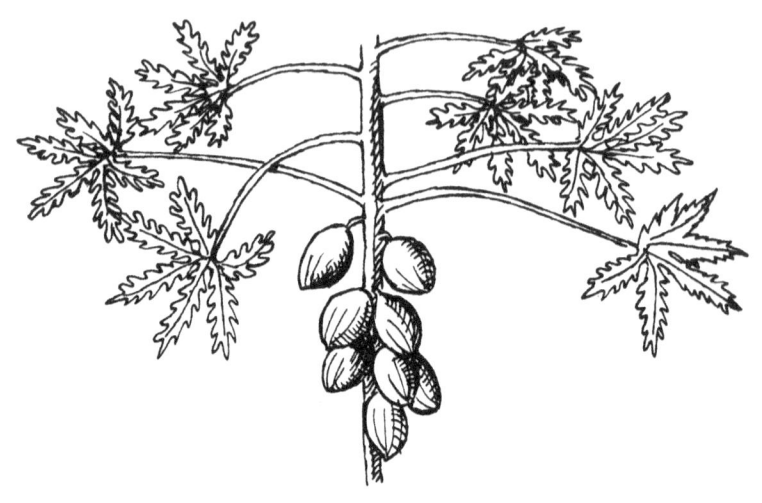

Papayas

The papaya, or "tree melon" as it is often called, is native to the American tropics. It was grown by the Aztec, Inca, and Mayan Indians of Mexico and South America for centuries before the discovery of the New World.

Early Spanish and Portuguese voyagers liked the fruit, and carried its seeds with them to many points around the world. By 1650 it was being cultivated extensively in parts of Asia, Africa, the Mediterranean area, and throughout the South Pacific Islands.

Papayas were introduced into Florida by early Spanish settlers. Because of the tropical climate required for their growth, their production in this country is still limited chiefly to that state.

The word "papaya" is a corruption of the Carib Indian name "ababai." It is also sometimes called "papaw" or "pawpaw" in English, "fruta bomba" in Cuba, and some of the other Spanish-speaking countries, "milikane" and "he-i" in Hawaii, and "iita" in Tahiti.

The papaya grows on a giant tree-like, herbaceous plant. The fruits range from one to twenty pounds each, and mature about 18 months from the time the seed is planted. They are shaped like an oblong melon, and have smooth, thin skin which is green when immature and yellow to orange when ripe. The flesh is deep yellow to orange, and has a sweet delicate flavor. A large central cavity contains the mass of small dark seeds.

Papayas are highly popular throughout the tropical world, and have long been a staple food in Asia and Africa. While still not well known in the United States, they are beginning to be shipped more extensively, and are slowly gaining recognition.

PAPAIN is a by-product of the papaya. It is an enzyme, or organic substance, which aids in the digestion of foods.

The West Indian peoples, as well as many others, have long been familiar with this property, and use the leaves of the papaya tree around pieces of meat while cooking, to make it tender. Others add lumps of the green fruit to their meat stews.

Dried papain is the base of many of our common meat tenderizers.

Peaches

Peach pits dropped into the sewing basket of the mistress of a Georgia plantation brought America one of its favorite peaches. It began when a gentleman in Delaware sent peach tree buddings to his friend, Samuel Rumph, of Marshalville, Georgia, in 1857. The trees flourished and Mrs. Rumph accidentally dropped a few pits into her sewing basket. Some ten years later, when her young grandson was starting an orchard of his own, Mrs. Rumph dug out the pits, dried and old, and Samuel planted them.

An accidental cross-pollination took place, a casual miracle of the wind and the bees, and in 1870 Samuel's orchard produced a spectacular golden peach, a species brand new in the fruit world. Samuel named it for his wife, Elberta, and was a pioneer in packaging his product attractively and working out refrigeration for shipping.

Even though William Penn found the Indians cultivating peaches in their gardens, the peach is far from being a native American. It had to travel around the world for nearly 4,000 years before it settled here. The story of the wandering peach is the story of the migrations of people from Southeast Asia, to their contacts with China, India, and Egypt.

For some 2,000 years the peach was believed to have come from Persia, and the fruit was known as "Persian apples." But curious botanists have now traced it to China where it is mentioned in literature of about 2000 B.C., whereas Sanskrit and early Hebrew writings do not mention it at all, indicating that it was unknown in Persia in 1500 B.C. Virgil was the first Roman to write of peaches in the 1st century before Christ, but they are thought to have reached Greece 300 years earlier.

Peaches, like apples, came to this country with the earliest settlers. The first probably came to Mexico and were carried north to be planted in California missions and spread by Indians of the Southwest. The English brought them to Jamestown, Virginia, and Massachusetts, and they came with the French to Louisiana. Three varieties were growing in Delaware in such profusion that William Penn advised women to dry them.

The peaches we enjoy today were all developed in this country, some by happy accident like the Elberta, others by intense effort. A New Jersey experimental station once pollinated 30,000 blossoms in a search for new varieties, and the results were such juicy fruit as the Golden Jubilee and the Cumberland Experimental.

The West Coast regions pour a rosy harvest across the nation today, but a little more than 100 years ago hungry gold seekers in San Francisco paid $1,350 for the crop of a single tree!

Other fruitful areas are in the East and parts of the South and in the Great Lakes region. When it is harvest time in Romeo, Michigan, busloads of visitors are taken to the orchards to eat their fill of pink and gold tree-ripened fruit.

John R. Magness in *National Geographic* calls the peach the "most versatile of fruit," pointing out that it "can be eaten whole like an apple, sliced with cream, dried, stewed, pickled, spiced, canned, distilled into a fine liqueur, cooked into pie or jam — or frozen into delicious ice cream."

Pears

Pears were so popular in France around 1850 that they were often mentioned in songs and in verse. Raising prize specimens was a fashionable hobby.

The pear, in fact, has been one of man's favorite fruits for as far back as 1000 B.C. The ancient Greek poet Homer, in his *Odyssey*, listed pears as growing in the garden of Alcinous, King of the Phaeacians.

Theophrastus, Greek "Father of Botany" (37–287 B.C.), mentioned both wild and cultivated pears, and Pliny, Roman author (A.D. 23–79), named more than 40 varieties.

Early pears were wild. Later they were grafted and cultivated. The Romans are said to have distributed the fruit throughout temperate Europe, and at the time of the discovery of America many varieties were known in France, Italy, Germany, and England. Earliest settlers brought them with them to the New World. They were introduced into California by the Franciscan Fathers, who planted them at their missions.

Today, more than 3,000 species are known. However, only about 20 of these are popular. They may be divided into two general classes: the "European" type pear, which has a soft consistency, and the "Asian" type, or "sand pear," which has hard flesh with gritty cells.

Pears are one of the few fruits that do not ripen well on the tree, and are therefore usually picked green. Ancient Chinese ripened green pears by putting them in a closed room with burning incense.

Persimmons

Has anyone ever coaxed you to bite into a green persimmon? If the answer is yes to this age-old prank, then you are familiar with the unpleasant puckery sensation. This is due to the presence of an acid in the unripe fruit called tannin.

Though singularly inedible when green, a soft, ripe persimmon is one of our most delightful fruits.

Two kinds of persimmons are grown in the United States: the American persimmon and the Japanese or Oriental persimmon.

Our native species grows throughout the southwest quarter of the country, and a few trees are found in southern New England and Michigan.

Early colonists and settlers were delighted with the new-found fruit. Captain John Smith referred to it in his letters as "Putchamis" and noted that although it was torment to the mouth when green, it was as delicious as an "apricock" when ripe.

The Japanese persimmon, or "kaki" as it is known in Japan, should more properly be called "Oriental persimmon" since there is good reason to believe it is native to

China. It was imported into Japan from that country centuries ago.

Commodore Perry brought Oriental persimmons back to the United States with him at the time of his famous expedition in 1852, which opened Japan to world commerce.

There are many varieties of the Oriental persimmon throughout the islands of Japan and parts of China, where it enjoys great popularity.

Persimmons grow best in a subtropical climate which is neither too hot nor too cold.

Most varieties are astringent and puckery until dead ripe, although there are some, especially those having dark-colored flesh, which are mild and non-astringent. In many parts of the Orient green persimmons are placed in casks and sealed tightly to remove the "pucker."

In addition to being delicious out of hand, persimmons may be served cold with cream, or in puddings and cakes.

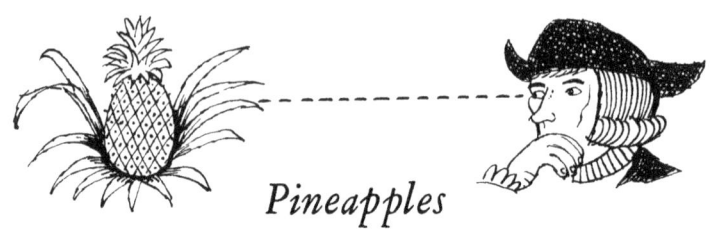

Pineapples

The pineapple is believed to have originated in Brazil and Paraguay, and is one of the most esteemed of tropical fruits.

The Guaraní Indians of Paraguay took it with them when they spread to northern South America.

"Anana," as pineapple is called in many countries, is from the Guaraní "*a*," meaning "fruit," and "nana," meaning "excellent."

When Columbus made his second voyage to the New World in 1493 he found pineapples growing on the island of Guadeloupe in the West Indies.

Early Spanish conquistadores called it "piña" because of its pine-cone shape. The English, also for this reason, called it "pine" (apple).

Pineapples grow on a large plant with long, stiff, grasslike leaves. The fruit is generally seedless, and new plants are produced by planting the shoots of the mother plant.

Although this fruit grows in most tropical countries today, the United States receives the bulk of its supply from Hawaii, Mexico, and the Caribbean Islands.

There is a great difference in flavor between the fruit which is ripened on the plant and that which is picked green, for shipping purposes. The reason for this is that there is a large amount of starch stored in the stem of the plant which turns to sugar during the last stage of the ripening process. When pineapples are picked green before this occurs, they are deprived of this sweetener.

Pineapple is delicious fresh, canned, or frozen. To prepare fresh pineapple, trim off the spikes and pare, beginning at the base. Remove all the outer part, dig out the "eyes" with a knife point, and slice. Discard the hard center core.

Plums and Fresh Prunes

The story of the American plum is the story of Luther Burbank, the mild-mannered New Englander who developed a new variety of potato (now named for him) on his Massachusetts farm and took his earnings of $150 to take a trip to California. Delighted with the sun and rich soil, he settled in Santa Rosa, where he worked hundreds of plant-breeding miracles.

Luther Burbank died in 1926, at the age of seventy-seven, but his tools still stand in his greenhouse and he is buried nearby under his favorite tree, a deodar, "divine tree of the gods," sent to him from the Himalayas. Behind him he left a rich heritage of fruits, vegetables, nuts, grasses, ferns, and flowers. He is credited with some 60 varieties of plums, including the Burbank, Climax, Wickson, Gold, and Apple plums and three new varieties of prunes.

Plums are the most widely distributed of all native fruits in America, and were a welcome and valuable addition to the diet of early pioneers. The Indians had gathered and dried the fruit for centuries before the arrival of the white man.

The most important of the plums grown in the United States today, however, came to us from Southwest Asia by way of Europe.

Earliest writings of European plums date back over 2,000 years. Pits of the Damson type have been found in the Lake dwellings of Switzerland, and Greek poets of the 6th century B.C. mentioned it. Pliny the Elder, Roman author, described several kinds of plums.

The "Greengage" or "Reine Claude" group, which are round, and range in color from light green to golden yellow, derived their name from Queen Claudia, wife of Francis I of France. They were introduced into that country around 1500, during the rule of her husband, and were named in her honor. Sir William Gage took them to England somewhat later, whence they were given the name "Greengage" in that country.

The Damson plum, so highly prized for making jam and plum butter, was named for the city of Damascus, capital of Syria.

English colonists brought pits of the various European plums to the New World for planting in their colonies. The French likewise took them to Canada.

The Japanese plum was introduced into California about 1870, and today it is another of our popular varieties. Actually, it is native to China, and reached Japan from that country some 300 or 400 years ago.

Plums are delicious eaten raw, and the **prune**-type varieties are dried and shipped throughout the world in large quantities.

Pomegranates

Grenadine, the popular flavoring for drinks, is made from pomegranate juice, while the tough rind of the fruit is used for tanning and making indelible ink.

The pomegranate, often called "the fruit of the Ancients," gets its name from the Latin, meaning "apple with many seeds" and is native to Persia (today Iran) and adjoining areas. It was frequently mentioned in the Old Testament of the Bible, in Sanskrit, the ancient literary language of India, and Homer wrote of it in his classic *Odyssey*. It was also pictured in Phoenician and Carthaginian medals, and depicted in ancient Assyrian and Egyptian sculptures.

Pomegranates grow on a small, bush-like tree. They are thick skinned, about the size of an orange, and range in color from pink to bright red. Inside, the fruit is divided into cells, which contain numerous seeds, each encased in a crimson, juicy pulp.

Pomegranates were brought to the New World by early settlers who planted them in California and what is today the Gulf States. These are still the two major centers of production in this country.

To eat the fruit raw, break open, dislodge the kernels, and suck the flesh from the small pits. A delicious and refreshing drink is also made by crushing and straining the pulp.

Quince

The quince is closely related to the pear. It is native to Persia (today Iran), Greece, and the surrounding area.

The Greeks grafted a common variety of quince with one from Cydon in Crete, from which the present name of the fruit was derived.

There were many ancient legends surrounding the quince. It appears in Greek statues, and was represented on the walls of Pompeii. Pliny the Elder, Roman author and naturalist, wrote of a belief that the fruit of the quince "warded off the influence of the evil eye."

Quinces grow on a shrub, or small gnarled tree, which belongs to the rose family. The fruit is bitter and astringent when raw, but has long been a favorite in the preparation of jellies and preserves. It is also tasty cooked with other fruits.

Early colonists brought the quince to the New World.

The Japanese, or flowering quince, is grown for its beautiful and colorful blossoms.

Raspberries

The raspberry, along with other berries, has been enjoyed by man from earliest times.

It is a bramble bush, of the rose family, and is native to the temperate zones of North America, Europe, and Asia.

Pliny was among the first to mention raspberries in his agricultural writings. He stated that they had been introduced into Rome from Mount Ida in Greece.

The fruit was relatively unimportant in Europe for many centuries. It was mentioned briefly in the 17th century in a book on orcharding, but it wasn't until the 19th century that an attempt was made to improve the size and quality of the berries through cultivation and breeding.

Early American colonists found raspberries and other berries growing in abundance in the new land. Berries, in fact, were so plentiful that more often than not they were plowed under to clear the ground for farming.

Toward the middle of the 19th century a lively interest in fruit breeding was awakened in America, and some of the fine varieties of raspberries known to us today were developed at that time.

One of these, the Fannie Heath Raspberry, is a tribute

to a determined pioneer woman who came to North Dakota from Minnesota as a bride in 1881. She found that the tree claim of her homestead was on barren alkaline soil. Fannie, however, had set her heart on a flower garden, shade trees, fruits, and vegetables. Neighbors said it couldn't be done, but 40 years later her house was circled by box elder, cottonwood, ash and black walnut trees, and her flower garden boasted damask roses brought to Virginia in 1774 by another pioneer. When she died in 1931, the black raspberry she had developed on her bleak tree claim was given her name.

Variations of raspberries include black, red, purple, yellow, and pink.

Whether cultivated or gathered wild in your favorite berry-picking spot, raspberries are always a delicate treat.

Strawberries

The world's biggest strawberry shortcake is claimed by Lebanon, Oregon. A feature of the town's festival of strawberries and roses, it towers twelve feet in the air and has to be cut with a saw!

This big, juicy Oregon strawberry is typical of the cultivated berry so popular in America, and it is the product of a union between the wild strawberries of North and South America. Most of the improved varieties date from the 1890's although strawberry patches were common in the United States in the early 1800's.

The first fruit variety of any kind originated by a plant breeder in the United States was the Hovey which came from a cross-pollination made in 1838 by Charles Hovey of Cambridge, Massachusetts. Now the federal government and many state experimental stations are busy turning out superior new varieties from selected and crossed berries.

Wild strawberries were mentioned in Virgil, Roman poet (70–19 B.C.), and Pliny the Elder, Roman author, wrote of them.

Species of strawberry are native to most temperate zones

around the world. When the colonists landed in eastern North America they found a great abundance of wild strawberries. Samples of these plants were taken to France in 1624, according to Jean Rodin, gardener to King Louis XIII, and from there to England and other European countries. These berries, however, remained small, even under cultivation.

In 1712 a Frenchman, Captain Frezier, found strawberries growing in Chile in South America, which he described as being "as large as walnuts." The Indians in that section had been cultivating the berries for centuries before the arrival of the Spanish explorers.

Captain Frezier took some of these strawberry plants back to France with him. The crossing of the species from the two Americas produced the ancestors of our modern varieties.

Because they are easy to grow, strawberries are a favorite in home gardens. Commercially, they are one of our most important fruit crops.

Tangerines

Tangerines belong to the mandarin group of oranges, and like other citrus fruits have been cultivated in Southeast Asia since ancient times.

In general, tangerines have loose skin that separates easily from the flesh.

Mandarins are favored in Japan and China over other types of oranges.

They are said to have reached Europe about 1805, and soon became well known and liked in the Mediterranean countries.

Authorities in citrus fruits, state that mandarins were introduced into the United States around 1840, when the Italian consul at New Orleans brought some in and planted them on the grounds of the consulate. From there they were taken to Florida and California.

The name "tangerine" is said to have been derived from Tangier, Morocco.

Tangerines are an ideal fruit for picnic and school lunches.

Watermelons

In the United States we think of the watermelon almost entirely as a dessert, but in many arid districts of the world it is an important source of water to the natives. In others, it is a staple food and livestock feed.

The watermelon belongs to the large gourd family, which includes muskmelons, cucumbers, and squash. Its culture goes back to prehistoric times.

The ancient Egyptians prized the watermelon, and there are names for it in many of the old languages, including Arabic and Sanskrit. It has been grown extensively in the warmer parts of Russia and the Middle East for thousands of years.

Up until a little over a hundred years ago the watermelon was believed to be native to Asia; then Dr. David Livingstone, the great missionary-explorer (1813–1873), found large tracts of wild watermelons growing in Central Africa. Authorities have now established Africa as its place of origin.

The Moors introduced watermelons to southern Europe early in the Christian era, and it reached China about a thousand years ago.

Early European colonists took watermelon seeds to Brazil, the West Indies, the eastern North American mainland, and the islands of the Pacific.

There are many varieties of watermelons, ranging from very small ones to very large. The color of the pulp also varies, from pale pink and yellow to deep red.

Watermelon rind is a favorite for making pickles in America, and in parts of Asia the seeds are roasted and eaten from the hand.

VEGETABLES

Artichokes

The flower bud of an exotic thistle is still something of a gourmet vegetable, although those who fancy them would probably say that never in their lives have they eaten thistles. But they have; they call them artichokes.

Artichokes are harvested when the flower head is in bud. If it were to go on growing, the bud would burst into a feathery bloom of purplish white. Long ago, in Spain, these flowers were dried and used to curdle milk for cheese-making.

Americans prefer the familiar globe, or French artichoke. But another form called cardoon grew here 200 years ago, and long before that it had been a favorite of the ancient Greeks and Romans. Only its young leaves and immature stalks are eaten and these are grown in the dark to keep them white and tender. Roman housewives paid premium prices for the cardoon when they went to market in the second century after Christ.

Artichokes are native to the Western and Central Mediterranean lands, and were carried to Egypt and eastward over 2,500 years ago.

Our present form of artichoke, first recorded in Naples, Italy, about A.D. 1400, has a flower head, the base of the petals, and inner "heart" of which are the edible portion.

The name "artichoke" is from the northern Italian word "articiocco" and "articoclos."

French settlers in Louisiana are said to have brought it to the United States.

Artichokes require a cool, foggy climate, and the center of their production is in the Half Moon Bay district, south of San Francisco, California. This area has just such a climate.

Artichokes are boiled until the stem base is tender. They may be eaten either cold or hot. Each leaf or petal of the bud is pulled off, and dipped in the accompanying sauce, and the tender end of the leaf only is eaten. When the leaves have all been stripped off, the base or "heart" is also a delicious morsel.

✿ ✿ ✿ ✿ ✿

JERUSALEM ARTICHOKES The Jerusalem artichoke did not come from Jerusalem . . . and it is not an artichoke! It belongs to the sunflower family, and is native to North America. In contrast to the prickly globe or French artichoke, it is a knobby tuber which grew wild across North America and was relished by the Indians. Eaten raw, the Jerusalem artichoke is crisp and succulent and is grown in small quantities for use in relishes and pickles.

Asparagus

Even before asparagus was used as food by man, it was considered a cure for almost anything from heart trouble to toothache. Its root is still used in many of our modern medicines.

Asparagus belongs to the lily family, and is believed to be native of the eastern Mediterranean lands and Asia Minor.

The Phoenicians, first "traveling salesmen," introduced it to the ancient Greeks, who used it in its wild form. The Romans cultivated asparagus as early as 200 B.C., and were so fond of it they dried the shoots for use even when out of season.

North Europeans and Britons have been eating asparagus for as long as there are records about it, and earliest voyagers brought it to the New World.

Today, asparagus is a universal favorite. There are approximately 150 species, some of which are grown for ornamental purposes, as the asparagus fern is very decorative.

It is a perennial plant, thrives near water, and under favorable conditions may remain productive for as long as 35 years.

To preserve the flavor of asparagus it should be cooked with a minimum of water, and as rapidly as possible. In this connection, the Roman Emperor Augustus, who is said to have been very fond of asparagus, originated the saying: "Quicker than you can cook asparagus."

Beets

Ancient Greeks valued beets so highly that they made small replicas of them in silver.

Garden beets, sugar beets, and Swiss chard belong to the same species. They are believed to have originated in the Mediterranean area and spread eastward in prehistoric times.

At first only the leaf of the beet or chard was eaten, and its small roots were used for medicinal purposes. The type of beet plant that produces a large edible root, such as we have today, was unknown before the Christian era.

Until 1800 the cultivation of beets remained relatively unimportant, but from then on interest in their propagation increased and they soon spread throughout Europe and the Americas.

Beets, and their leaves, have a high nutritional and vitamin content, and are a favorite in our markets today. They are also canned in large quantities, and during World War II were found to be one of the most satisfactory vegetables for dehydration and use by the army.

SUGAR BEETS, now cultivated extensively in the United States, were developed in France and Germany around 1810. Although they may be eaten as a vegetable, they are not so tender as the garden variety, and their main use is in making sugar. They are usually yellowish-white in color.

Sugar beets should not be grown near the edible red beet, because the wind-blown pollen will fertilize the pistillate flowers of both varieties and will produce curious cross-breeds, the white of the sugar beet paling the red beet and the latter bringing an unwanted blush to the sugar beet.

Broccoli

"Italian asparagus" appeared on British menus in the early 1700's, and the somewhat baffled but venturesome English tasted for the first time the tender green stalk and cluster of green buds which everyone now knows as broccoli.

Yet these delicate "cabbage flowers" are among the ancients of the vegetable world. Broccoli is native to the Mediterranean area and Asia Minor, and has been known in Europe for more than 2,000 years. According to Pliny, Roman naturalist of the first century A.D., 23–79, the Romans grew sprouting broccoli at that time, prizing it highly.

"Broccoli" is an Italian word from the Latin "brachium," which means "arm" or "branch." Like cauliflower, it is an aristocrat of the cabbage family, and is grown for its thick, undeveloped flowers and stalks, rather than for its leaves.

There are two general types of broccoli: the "heading" type with dense white buds like the cauliflower, and the "sprouting" kind, which forms branching clusters of green flowers on its tall green stalks, along which are smaller sprouts.

Broccoli was not introduced into England until around 1720, where it was called "Italian asparagus." Early Italian settlers brought it to the Americas, and although it was slow in gaining popularity here, today it is an important market and home-garden plant.

It is an excellent source of vitamins and minerals. To preserve these nutrients, it should always be cooked rapidly, with as little water as possible.

Brussels Sprouts

Brussels sprouts, like kohlrabi, are among the very few "new" vegetables, and were unknown until about 400 or 500 years ago.

The plant is a member of the wild cabbage family, and consists of a tall stem with many tiny individual heads of leaves, rather than one large single head.

It was developed in the vicinity of Brussels, Belgium, from which it gets its name, and was first mentioned by a European botanist in 1554. It was not known in America until around 1800.

Brussels sprouts are grown for the fresh-vegetable market, and also for freezing and canning.

They contain all of the valuable nutrients of the cabbage family. When cooking, a dash of nutmeg will add an exotic and unusual flavor.

Like all members of the cabbage family, Brussels sprouts will hybridize with no inhibitions. As a result, growers producing commercial seed must be on guard to prevent disastrous hybrids of cabbage with kale, cauliflower, kohlrabi, broccoli, and collards.

Cabbage

A New England farmer looked over his strawberry patch and shook his head. The berries were small, the plants puny. Then he looked at the cabbages growing next door and shook his head again. The cabbages looked terrible. The farmer didn't know why. But practitioners of a new technique, the Bio-Dynamic Method of farming, knew exactly what was wrong: Cabbage doesn't *like* strawberries.

But surround cabbage with blossomy, aromatic plants and the cabbage will emerge green, firm, and healthy. Experts say this is because cabbage is the victim of a sort of one-sided evolution in which the flowering process is subordinated to the enormous terminal bud or cabbage head. This weakens the cabbage and makes it dependent upon "help" from other plants. Good companions have been found to be sage, rosemary, hyssop, thyme, and peppermint.

Cabbage is believed to be the most ancient of all vegetables grown today, and has been cultivated for more than 4,000 years.

It was worshiped on elaborate altars by ancient Egyptians, who considered it a god. It is also mentioned in Greek mythology as having "sprung from the sweat of Jupiter" while he was trying to reconcile two oracles; and early medicine men recommended it as a cure for many ailments, including baldness.

Cabbage is a member of the large mustard family, but belongs to a distinct separate branch which includes cauliflower, kale, broccoli, and Brussels sprouts.

The name "cabbage" is applied in general to types of this plant with inner leaves which form a center head. Early cabbage, however, was non-heading and long-stemmed.

Because of its antiquity the derivation of its name is somewhat obscured, but authorities generally concur that it comes from the Latin word "caput," meaning "head." It is believed to have been introduced into Europe by the Celts at the time of their invasions around 600 B.C. but did not reach the Orient until much later, as indicated by the lack of any name for it in Sanskrit or other ancient languages of the East.

Among other attributes, ancient Greeks and Romans said that cabbage counteracted the effects of alcohol, and Aristotle, the Greek philosopher, and later Roman statesmen, dined on cabbage or cole before attending a banquet at which much wine was to be served.

Cabbage was brought to the New World by Jacques Cartier, the French navigator who discovered the St. Lawrence River in 1536. It was planted by early settlers and Indians, and quickly became an important food staple.

There are many types of cabbage, but they may be divided into two general groups: the "loose-heading" and the "hard-heading." They vary in colors from dark green and red to white.

Cabbage is a rich source of all of the important nutrients.

Carrots

When carrots were first brought to England, ladies loved their feathery leaves, and used them to decorate their hair.

The carrot has been known since ancient times, but for many centuries was grown for its medicinal value only. Ancient Greek physicians recommended its use, especially as a stomach tonic.

It is believed to have originated in Afghanistan, and gets its name from the Latin word "carota."

Carrots did not become generally popular as a food until the 13th century, at which time they were being raised throughout most of Europe. They were also known in China, where they were taken from Persia.

European voyagers carried carrots to the New World, where they were quickly distributed throughout the Americas. The Indians adopted carrots enthusiastically, and planted fields of them.

There are many varieties of carrots, but the most popular in the United States is called the "Mediterranean" type. It is long, and orange colored. Other forms are round, and colors vary from red and purple to white. The "Japanese carrot," raised in the Far East, is commonly three feet long or more.

As did the ancient Greeks, we enjoy carrots for good health today. They are one of our most valuable sources of Vitamin A.

To remove the skin from carrots easily, simply drop them in boiling water for five minutes, then in cold, and slip off the skins.

Cauliflower

Mark Twain is said to have described cauliflower as the "cabbage with a college education." It is indeed, along with broccoli, an aristocrat in the large cabbage family.

Cauliflower derives its name from the Latin "caulis" (cabbage) and "floris" (flower).

The oldest record of cauliflower dates back to the 6th century B.C. Pliny, Roman naturalist, wrote about it in the 1st century A.D. It was introduced into Spain from Syria in the 12th century, where it was said to have grown for over 1,000 years. It has also been known in Egypt and Turkey for about 2,000 years.

Cauliflower has long been a favorite throughout Europe, and early settlers, principally from Italy, brought it to the Americas. Because of its delicate flavor and dietary value, it is an important item on our American menu today.

Cauliflower is easily grown in home gardens, and matures in three to five months.

In addition to its many uses as a cooked vegetable, it is used extensively for making pickles.

Celery

In ancient Greece a stalk of celery was handed to the winner of an athletic event. Such was the great esteem in which this plant was held.

Celery, or "smallage," as it was called in its wild form, originated in the Mediterranean countries, and has been known in that area for thousands of years. It is said to be the "selinon" mentioned by Homer in his *Odyssey* about 850 B.C.

Our word "celery" comes from the French "céleri," which is derived from ancient Greek.

It belongs to the parsley family, along with carrots and parsnips. The distinctive flavor and fragrance of these plants are due to vaporous oils contained in their stems, leaves, and especially in their seeds.

Ancient celery, in its wild form, was bitter, and was used only as a medicinal plant as late as the 16th century. It was not until the late 17th century that it was cultivated as a food in France, England, and Italy. Chinese writings of the 5th century A.D. also mention its medicinal values.

Early colonists brought celery to the New World. One night in the 1850's, guests in the Burdick House, Kalamazoo, Michigan, encountered a dish of strange-looking green stalks on the supper table. A venturesome few tried it, but no one was enthusiastic. James Taylor, a Scottish immigrant who had grown the green celery in his backyard, was disappointed. He had been sure that the crisp green variety was better than the pale, bleached celery known on the East Coast.

Mr. Taylor gave up, but not so a nearby farmer named DeBruin. He, too, began growing celery and sending his large family of children to sell it door to door. Today Kalamazoo is nationally famous for celery.

The idea of growing celery on the West Coast originated in 1891 in the marshlands of Orange County, California. A farmer tried recruiting local residents, who were known as "tule-rooters and swamp angels." When his first crop failed, he thought about the Chinese market gardeners. They had the necessary skill and by 1893 thousands of acres were in celery culture and the marshland had rocketed from $15 to $400 an acre. By an ironic paradox, these creators of a million-dollar industry were so resented by local residents that armed guards had to be posted around the celery fields; another chapter in the eventful forward march of vegetables in the United States.

* * * * *

CELERIAC, a large-rooted variety, more popular in Europe than in the Americas, is not used raw, but is well suited for soups and cooked dishes.

SMALLAGE, the original "wild" celery, is still cultivated extensively for seasoning.

Crisp celery is an ideal addition to any meal, or for an in-between snack, especially for dieters, as it is nutritious but very low in calories.

Chicory

"Black as sin, sweet as love" is a favorite description of New Orleans coffee . . . and chicory is what makes it like that.

Chicory is closely related to endive-escarole. It has long been raised as a garden plant in Europe, and now grows profusely throughout the Temperate Zone.

Its leaves are good in salads, and as a potherb. The roots, when young, are cooked much the same as carrots. Dried, roasted, and ground, they are used as an adulterant, or substitute for coffee.

Collards

Collards, like kale, are among the most primitive cultivated members of the cabbage family.

The name comes from the Anglo-Saxon "colewarts," meaning "cabbage plants."

Because of their antiquity and widespread use, it is difficult to determine the exact place of origin of collards. Some authorities place it in the eastern Mediterranean area, while others maintain it originated in Asia Minor.

The Greeks grew collards along with kale, and made little distinction between them. They were also popular with the Romans, who carried them to Britain and Europe in the wake of the earlier Celt propagation.

Nutrition experts today highly recommend the inclusion of both kale and collards in our diet, because of their rich vitamin and mineral content.

Collards may be obtained throughout the United States, but are especially popular in the Southern states. They are also a favorite for home gardens, because they are easy to grow.

Corn (Maize)

Corn, one of the great food gifts of the New World, is also a tantalizing mystery. No one knows where or when two grasses were accidentally crossed to become the ancestor of Indian corn, whose proper name is maize. Serious students of botany range the Andes and the interior of Guatemala seeking a clue to the origin of corn.

The most helpless of plants, corn cannot travel by itself. It must be carried by man, carefully cultivated by him. Even in the man-planted field, "volunteer" corn is a rarity.

Yet when the white man arrived in the Americas, corn was growing across thousands of miles, from South through Central America and North America. Indian tribes who apparently never ranged far from home territories were cultivating dozens of varieties of corn. Someone, sometime, must have brought the first plants there.

Without corn the colonization of the New World might never have taken place. Remember Squanto bringing Indian corn to the starving Puritan colony? Leif Ericsson, believed to have landed south of Cape Cod in the year 1000, reported sown fields of what he called wheat, but what we know to have been corn. The Mayan and Incan empires were, literally, built on corn. Montezuma accepted corn as rent, and as late as 1671, freemen of Martha's Vineyard in Massachusetts used corn and beans for votes — "the corn to manifest election, the beanes contrary."

Indians of our Southwest had perfected irrigation for their cornfields thousands of years ago, and botanists now believe they had also learned to hybridize corn.

So important is the miracle of corn that Indians throughout the Americas held it sacred, a gift from the gods. Each tribe had its own legends and rituals centering around corn. The Jemez tribe in New Mexico believes corn was born there, from a tasseled grass still growing in Frijoles Canyon. The Navajos tell of a great turkey hen flying overhead many, many moons ago and dropping an ear of blue corn on the land. The Iroquois legend is of a spirit woman walking the fields with corn and pumpkins growing in her wake.

Zuni medicine men sprinkled corn across trails to prevent the Spanish Conquistadores from entering tribal villages. Creek Indians held a harvest thanksgiving ceremony. Before the great day old pottery was broken, hearths and homes cleaned, old clothes thrown away and a new fire kindled in honor of the ripened corn.

The Indians soon established the magic link between corn and livestock when the Spanish gold seekers brought in cattle and hogs, a relationship that has come to fabulous fruition in the corn-hog economy of the Middle West. The five states of Ohio, Illinois, Indiana, Missouri, and Iowa produce more than a billion and a half bushels of corn every year... nearly half the nation's total production, and more than one-fourth of the world production! The soil and climate are ideal in the Midwest for this vital feed crop. On hot August nights, when the corn wind blows, farmers will listen for a soft, secret sound in the dark. They will nod with satisfaction, because it means a good harvest when "you can hear the corn grow."

South Dakota pays tribute to corn in a six-day festival held in the world's only Corn Palace, in the city of Mitchell. Some 2,000 bushels of corn are made into pictorial panels.

One season the "pictures" told the story of the state from an Indian roasting buffalo over a prairie fire to a tractor standing beside the sweeping fields of a streamlined modern farm.

Sweet corn, as we know it today, is golden yellow, creamy white, or greenish. But once it came in many colors, and in native markets of Mexico, South and Central America you can still find ears of corn in blue, black, red, or particolors.

While the Incas, Mayas, and Aztecs used corn for food, fuel, currency, jewelry, smoking silk and building material, today we have found more than 600 uses for it! These range from bourbon whiskey to sizing for slick magazines ... and of course that fun member of the maize family, popcorn.

Cowpeas

George Washington was a fancier of cowpeas, and records show he bought 40 bushels of seed for sowing on his plantation in 1797.

The cowpea originated in India, and reference to it in the ancient Sanskrit language indicates its cultivation from earliest times.

From India it was carried to Arabia and Asia Minor, and thence down into Africa in prehistoric times.

The name "cowpea" is of American origin. Despite this designation, however, it is distinctly different from the "English" or "garden" green peas, and botanically is closer to the bean family.

Most of our first garden plants reached America by way of Europe, but cowpeas arrived from the island of Jamaica, where they were brought from Africa by slave traders around 1675.

Cowpeas thrive in a warm climate, and from early colonial days have been an important food throughout our Southern states.

They are harvested when the seeds are formed, but before the pods dry out. The "peas" are then shelled out. They are highly nutritious, with a fine savor all their own, especially when prepared "Southern style" with a bit of ham or pork fat.

There are numerous varieties of cowpeas, but the most popular is one called "Black-Eye."

TIBERIUS

Cucumbers

"Cool as a cucumber" is not just a catchy phrase. A thermometer reading of the inside of a field cucumber on a warm day registers approximately 20 degrees cooler than the outside air. Cucumbers enjoy shade and get on very well in a cornfield.

The word "cucumber" comes from the Latin name "cucumis," and it is one of the four important vegetables given to the world by India. The others are cowpeas, Indian mustard, and eggplant.

The cucumber is of great antiquity, and ancient literature indicates its culture in western Asia is over 3,000 years old. It is one of the few vegetables mentioned in the Bible: in *Isaiah 1:8* "a garden of *cucumbers*," and again in *Numbers 11:5*, along with melons, leeks, onions, and garlic.

Cucumbers were introduced into China from Persia around 100 B.C. by a Chinese ambassador who found them in his travels.

They were known to ancient Greeks and Romans, and the Roman Emperor Tiberius is said to have ordered cucumbers served to him every day of the year, which required artificial forced culture to grow them out of season.

Charlemagne, King of the Franks and Roman Emperor of the West, had cucumbers grown in his gardens in France in the 9th century.

Columbus brought cucumbers to the New World, and today they are a universal favorite.

There are many varieties of cucumbers, ranging from the very small gherkins to the great English greenhouse kinds that are often nearly two feet long.

In the United States, cucumbers are used primarily in raw salads, and of course are a favorite pickled.

Eggplant

Something about the eggplant, its deep purple-black color, its rounded shape, led the painter Charles Bemuth to make a long study in sketches and on canvas of this vegetable. His major "Eggplant" canvas, glowing with color, has been shown in museums across the country.

But for a long time the eggplant was as suspect as the tomato. Some northern European botanists called it the "mad apple" and warned that it could induce insanity, while in 16th century Spain, quite the opposite view was taken and the eggplant was sought after as the "apple of love."

Eggplant is believed to have originated in India, and was also known in China and adjacent areas in prehistoric times. It appears to have been unknown in the Western world until about 1,500 years ago.

There are numerous names for it in the ancient Sanskrit and Hindustani languages, but none in the ancient Greek and Roman.

One of the oldest records of eggplant is in a Chinese book written in the 5th century A.D.

Early Spanish settlers took eggplant to South America, and English colonists brought it to North America.

There are many varieties of eggplant today, ranging from very small ones to the very large. They come in a large assortment of colors, from deep purple to light yellow. Striped forms are also known. The most popular is the large purple sort.

Eggplant is universally known today, but is especially important as a vegetable in the Orient.

Endive-Escarole

Endive-escarole is native to the East Indies. The histories of Egypt and Greece show that this vegetable was popular in those countries long before the Christian era. It was also used extensively by the Romans, both as a salad and cooked.

Its popularity spread into Europe through the ages, and early colonists brought it to the New World.

There are many varieties of the green, but it may generally be classed in two groups: the narrow curled-leaf type and the broad-leaf kind generally known as "escarole." French and Belgian endive belong to this latter group. They are usually imported from those countries, as they are not commonly raised in the United States.

Curly-leaf endive is a tasty addition to salads, or cooked as greens.

For a treat that is different, try the French or Belgian endive filled with mashed avocado to which a dash of salt and lemon juice have been added. Each broad blade is stuffed in the same way as "stuffed celery."

Garden Peppers

In the mountains of Tennessee there is a superstition that if a farmer plants peppers while he is angry at his wife, the crop will fail.

The "garden" or "sweet" pepper is not related to our common table condiment pepper, which comes from the dried and ground berries of a climbing shrub.

Sweet peppers are native to tropical America, where all of the numerous varieties were used extensively by the Indians centuries before the arrival of Columbus. Fragments of peppers have been found in ruins in Peru which are over 2,000 years old.

One of the principal objectives of Columbus and his associates was to find a shorter trade route to India and the East Indies, and among the most valued and important products sought from these countries were spices. When they discovered the pungent varieties of the "garden" pepper in the West Indies, therefore, they were elated to have found what they believed to be a new variety of the "table" pepper condiment. They took many samples back to Europe with them, where the new "find" was received with intense interest. The enthusiastic cultivation and propagation of the plant, plus the detailed records that were kept of its progress, are unique in the history of American plants.

Later travelers to the New World found a wide variety of species of garden peppers throughout the West Indies, Central America, Mexico, Peru, and Chile. Some of the hot kinds from Chile are called "chili" after the country, and in fact, "chili con carne," which means "chili (pep-

pers) with meat," is a favorite today throughout southwestern sections of the United States.

The pimiento, too, had to become a transatlantic traveler to reach the United States. It went with Columbus back to Europe and was immediately taken up by the Spanish. It was such big business that we imported practically all our pimientos from Spain until an enterprising Georgian brought in a few packets of seeds and began experimenting. Now the pimiento is back home and both growing and canning are done in this country.

Another variety, bright red and mild, has long been cultivated in Hungary and adjacent areas, and is ground to make paprika. Cayenne pepper is still another form of dried ground red peppers.

It is interesting that although the garden pepper is a native of the Americas, it was not until after it had become popular and well propagated throughout Europe, that it was later reintroduced into northern America. Many people in the United States still believe it to be a European vegetable.

Peppers contain important vitamins and minerals, and add much flavor and interest to our menu. They are a favorite stuffed with meat, rice, or sea food; a good addition to vegetable salads, and the sweet ripe ones are delicious cut in strips and baked with a little salt and butter until tender.

Garlic

Ancient Romans used garlic sparingly in their cooking, but believed it possessed magical powers, and fed it to their laborers to make them strong, and to their soldiers to make them courageous. They credited much of the success of their battles to the consumption of garlic.

Garlic is a pungent, strong-scented bulb, composed of smaller bulbs called "cloves." Its name comes from the Anglo-Saxon "garleac," "gar" meaning "spear," and "leac" meaning "leek." It is a member of the lily family, and its long history parallels that of the onion and leek.

From ancient times the medicinal qualities and seasoning attributes of garlic have been prized. Homer, Greek poet of the 9th century B.C., Pliny, the Roman author, and Aristotle, Greek philosopher, sung its praises in their writings.

It is said to have been introduced into China in the 2nd century B.C., and has been used as a seasoning and cure for many ills in India for centuries.

The Bible mentions garlic (*Numbers 11:5*) along with onions, leeks, cucumbers, and melons. It was popular with early Egyptians, and Europeans in the Mediterranean countries have used it extensively for over 2,000 years.

Early Spanish travelers brought garlic to the New World, and its culture and propagation were taken up enthusiastically throughout the Americas.

Good-quality garlic should be dry, with firm, well-shaped cloves. It should be kept in a cool, dry place.

There is no tastier seasoning than garlic, and a touch adds much flavor interest to meats, sauces, and vegetable salads.

Green Peas

Madame de Maintenon, secretly wed to King Louis XIV of France, wrote in 1696: "This subject of peas continues to absorb all others. Some ladies, even after having supped at the Royal Table, and well supped too, returning to their own homes, at the risk of suffering from indigestion, will again eat peas before going to bed. It is both a fashion and a madness."

The history of "green peas," also known as "garden" or "English" peas, is lost in antiquity, but their seeds have been found in lake mud excavations in Switzerland dating back 5,000 years to the Bronze Age. Other findings in Hungary date back even further.

Primitive eastern Aryans introduced peas to the Greeks and Romans, who grew them long before the Christian era, and our name "pea" comes indirectly from the Latin "pisum."

At first peas were grown only for their dry seeds, and "green peas" are not mentioned until after the Norman Conquest of England, where they were listed among other foods in the 12th century. No detailed description was given, however, until 1536 in France, and they were considered a great and very costly delicacy until the 18th century.

Before the end of the 16th century many kinds of peas were being developed in Germany, Belgium, and England, and they were brought to the New World by the earliest colonists.

Fresh green peas are not only one of our favorite vegetables, but are packed with valuable food energy. They are best when cooked very quickly, with as little water as possible. A leaf or two of chopped lettuce added to the peas while cooking gives them an especially fine flavor.

* * * * *

POD PEAS. One variety of peas, popular in parts of Europe and in the Orient, has a tender pod and very small peas. These are harvested when young and tender, and the entire pod and pea are eaten.

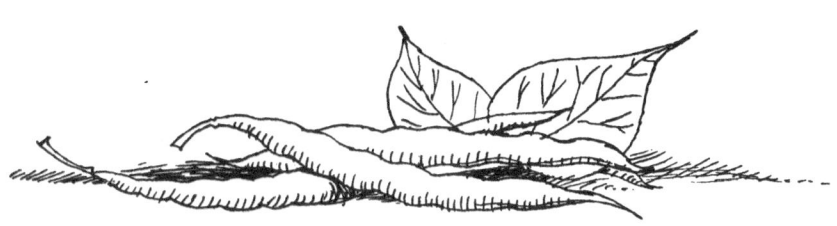

Green or Wax Snap Beans

Common beans, which include wax beans and string beans, are native to the Americas. The Old World was familiar with certain types of beans, but neither our "common" nor "lima" bean was known.

For centuries before Columbus discovered the New World, many types of beans had been cultivated by the Indians throughout the Americas, and their distribution has been traced along the prehistoric trade routes.

The Indians usually planted beans along with maize, or corn, as we commonly call it. Combined, these two vegetables supplied them with most of their food requirements, especially those tribes that used little or no meat.

Early explorers and captains of the slave boats of the 1500's found beans excellent for replenishing their ships' food supply. They obtained them from the Indians, and carried them to all parts of the world.

The word "bean" is not specific, and is used to refer to the seeds of many different kinds of plants. However, we understand it to mean one of our two general types: the "common bean," including wax beans, string beans, kidney beans, pinto beans, etc., and "lima" beans.

"Wax" or "snap" beans are cooked when fresh. Through many years of cultivation, the "string" bean is now actually "stringless."

"Field" or dry beans require different preparation, usually preceded by soaking them in water to soften the hulls.

Beans are one of the most important mainstays in the diet of millions of persons throughout the Americas and many other parts of the world as well.

Hominy

Hominy is the inner part of the maize, or corn grain, which has been soaked to remove the hull.

It is eaten whole, or ground coarsely into what is known as "hominy grits."

Hominy originated with the Algonquin Indian tribes of North America, who called it "rockahominy," from which we derive our name for it.

Hominy contains most of the food values of the whole maize kernel, and it is a great favorite throughout the southern and southwestern United States.

Kale

Kale, like collards, is one of the most primitive cultivated members of the cabbage family, and its history can be traced back over 4,000 years. It differs little today from the original nonheading "cabbage," used for food by man in prehistoric times.

Because of its antiquity and widespread use, it is difficult to determine its exact origin, but most authorities agree it was probably native to the Mediterranean area. From there it was gradually spread by the Celts and Romans.

Though it has many names, our word "kale" is Scottish, and is derived from the Greek and Roman "coles" and "caulis," which was their general term for the whole cabbage family.

The Greeks and Romans grew many varieties of the green, and European writers of the 1st, 3rd, and 4th centuries wrote of them all. Kale was brought to America by early settlers, who called it "colewarts."

Kale is one of our most common garden vegetables. It is very simple to grow, and is unusually rich in minerals and vitamins. Its tender young leaves add zest to salads, and it may be cooked in the same manner as cabbage.

Kohlrabi

Kohlrabi and Brussels sprouts are among the very few so-called "new" vegetables, and were unknown up until about 500 years ago.

Kohlrabi does, however, belong to the ancient wild cabbage family, but is different from the other members in that instead of a head of closely packed leaves it has a turnip-like enlargement of the stem just above the ground. This is the part for which it is grown, and from which it gets its name "kohlrabi," which in German means "cabbage turnip."

It was developed in northern Europe, and was first described by a European botanist in 1554. It was not mentioned in the United States until around 1800.

The "globes," which resemble white turnips in flavor, but are much more delicate, are very good when small and tender. They are usually steamed, peeled, and eaten hot, or cooled and served with French dressing as a salad. The tops are also eaten when young, and are cooked like other greens.

Kohlrabi is an ideal garden vegetable, as it is easy to grow and is remarkably productive.

Leeks and Chives

In the 6th century the Welsh attributed a victory over the Saxons to the leeks they wore by order of St. David to distinguish them in battle. It is to this day the floral emblem of Wales.

Leeks and chives are relatives of the onion and garlic. They are native to Middle Asia, with a secondary area of development along the Mediterranean, and have both been cultivated for food from prehistoric times.

The leek is similar to the onion, but has smaller bulbs and flat, succulent leaves. Its name comes from the Anglo-Saxon "leac."

Chaucer, English poet (1340–1400), wrote of leeks in his *Canterbury Tales:* "Wel loved he garleek, oynons, and eek lekes."

The Romans prized leeks, and Emperor Nero is said to have eaten great quantities, to the point of being nicknamed "Porrophagus" because of it. He believed, among other things, that they improved his voice.

They were also esteemed by the Egyptians, who used them from earliest times; and leeks are mentioned in the Bible (*Numbers 11:5*) along with onions, cucumbers, melons, and garlic.

The word "chive" is derived from the Latin "cepa," a form of onion. It has been grown for centuries in Europe and the British Isles, and is cultivated throughout the world today. It is also found wild in many localities, principally in Italy and Greece.

It has a purplish flower, and tubular leaves, which are used as a food seasoning.

Both leeks and chives are believed to have been brought to the New World by early colonists, although some authorities say that a kind of leek is native to this continent.

Leeks and chives add much interest to many of our salads and cooked dishes. The blanched stalks of leeks may also be prepared in the same manner as asparagus, and served with butter, or a cream sauce. And of course both greens are rich in all of the vitamins and minerals of the large "onion" family.

Lentils

The origin of lentils is lost in antiquity, but they are believed to be native to the Mediterranean area and Asia.

They are mentioned in the Old Testament of the Bible, *Genesis XXXV:34*, which tells how Esau, eldest son of Isaac and Rebecca, traded his birthright of priority to his younger brother Jacob, for a meal of "lentils."

They belong to the pea family, have a pale blue flower, and broad pods containing the edible seeds.

Lentils are highly nutritious, and are an important part of the entire food supply in many parts of the world.

They are delicious in soups, or as a vegetable, such as dried beans or peas.

Lettuce

Lettuce, one of the oldest food plants known to man, is undoubtedly the world's most popular salad green, and ranks high at the top of the list of every vitamin-conscious person.

It originated in India and Central Asia, and its culture has spread around the globe from ancient times.

Herodotus, Greek "Father of History," recorded that lettuce was served in the 6th century B.C. by the Persian kings. It was a favorite of Romans, who, because of its milky juice, called it "lactuca," from the Latin root word "lac," meaning "milk," from which our own word "lettuce" is derived.

Chinese of the 5th century also enjoyed it.

Rabelais, French author, is said to have taken it to France with him in 1537, and Columbus and other explorers brought it to the New World. Lettuce seeds were among the first planted by early colonists.

Although there are many varieties of lettuce, it may be

divided into two general groups: the solid "head" lettuce, in which the leaves form a cabbage-like head, and the "loose-leaf" species.

* * * * *

ROMAINE lettuce has long, straight leaves, and is so called because it was developed in Rome, Italy.

In addition to its important role in salads, it is a delicious addition to most other cooked vegetables.

California's Salinas Valley and vast irrigated acreage in Arizona now produce green miles of lettuce, still hand cut but then whisked by conveyor belt for quick trucking to market. Iceberg lettuce, said to have been developed at the urging of Aniello Crispo and other produce men for their customers at Washington Market in New York, and Boston lettuce, the loose-leafed, darker green type, are the two most popular kinds in the United States.

Lima Beans

Lima beans, it turns out, ought to be called Guatemala beans. They got the name we know them by because they were first taken to Europe by explorers who found them in Lima, Peru. These beans were part of several "bean migrations" which found our American varieties, "the common beans," traveling thousands of miles through Mexico to our Indian tribes.

Lima beans are fond of corn, as the Indians soon discovered. The cornstalk forms a natural pole for the beans to climb. The first New England settlers began following the Indian method of planting beans in the rows of corn and then of cooking them together in that classic American dish, succotash.

Lima beans, like our many varieties of "common beans," are native to the Americas, and were grown by the Indians for centuries before the arrival of Columbus.

Lima beans were at first thought to have originated in Brazil, but later findings point to Guatemala. Their distribution has been traced by the various prehistoric varieties left along old Indian trade routes, which spread throughout the Americas.

The name "Lima" came from the city Lima, capital of

Peru, where early European explorers found this type of bean. These travelers, and also the captains of the slave vessels, stocked their ships with great quantities of beans obtained from the Indians. Thus, they were carried to all parts of the world.

Lima beans are often called "butter beans" in some parts of the United States.

Manioc

Manioc, or "mandioca," as it is called in Spanish, is a staple starchy food similar to the potato. It is popular throughout tropical sections of Latin America, where it has been used extensively since earliest times.

It is the highly nutritious root of the cassava plant, and derives its name from the Tupian Indians of northern Brazil.

There is no history of manioc being popularized in Europe or the United States, and it is not common in our markets.

It is usually prepared by boiling until tender, and may be served with butter, or a gravy sauce. It is also used in making bread and tapioca.

Mushrooms

The mushroom is our "glamour" vegetable, and for centuries has been esteemed as a delicacy.

The pharaohs of Egypt, a thousand years before the Christian era, monopolized mushrooms for their own use, feeling that they were much too great a luxury to be eaten by the common people. They did not understand their sudden overnight appearance, and thought they grew magically.

Mayan Indians of Guatemala featured mushrooms in carvings in 1000 B.C., and the Romans called them "food of the gods."

For centuries only the wild types of mushrooms, found growing in the meadows and pastures, were known. Then, during the reign of Louis XIV of France, market gardeners of Paris began experimenting with mushroom culture. Soon they were cultivating them in abandoned caves from which building stone had been quarried, and in cellars, with much better results than outdoors. The British also raised mushrooms successfully in hothouses at that time, as did many other people.

The Plains Indians and Indians of the generally temperate parts of the United States knew mushrooms and used them in venison stews, in soups, and fried as a vegetable.

Although a favorite in the United States from colonial days, commercial production did not start until the latter part of the 19th century. It was climaxed in 1926 by the greatest event in mushroom culture history, when Lewis Downing of Downingtown, Pa., found a clump of white mushrooms among his usual cream-colored plants. Most of the mushrooms grown in the United States today are descendants of this white clump.

Mushrooms do not grow from seeds. They are a fungus, and reproduce by means of spores. These are tiny rootlike particles which spread through the organic matter in which mushrooms grow. Mushroom spawn is now cultivated by laboratory scientists who sell it to the growers for inoculation of the mushroom beds. Thus, where once one had to just hope for mushrooms to pop up, their production is now controlled.

Green plants get their food by manufacturing it in their leaves from air, water, sunshine, and soil nutrients, but mushrooms cannot do this. They have no leaves. So they must depend on green plants to make their food for them, and they cannot use it unless it is in the process of decay.

Thus, to cultivate mushrooms successfully it is necessary to have specially constructed, windowless sheds, with controlled temperature and humidity, and in addition great care must be given to the material in which they are grown.

When mushrooms first appear their tops are round and buttonlike. When mature, the cap flattens out like a parasol. They are generally picked at this state, although "button" mushrooms are also popular. Mushroom hunting has become an outdoor hobby and groups of people go out every weekend in season to find meadows and glens where the delicate caps are ripening. They are experts and they do not run the risk an amateur does when he attempts to tell the edible mushroom from its poisonous rival, the toadstool family. It has been only within the last century that botanists have succeeded in defining exactly which are edible and which are not.

The lovely "fairy ring champignon," which grows in circles just big enough for the "little people" to hold their moonlight dances, is a favorite on the continent, as is the Paris mushroom harvested from miles of carefully tended cellars.

Mushrooms are not only festive, but nutritive. They add zest to sauces and "dress up" almost any dish.

Indian Mustard

Indian mustard, which is usually simply called "mustard," is stated to have originated in northwest India, with a secondary area of development in Burma and China.

Our word "mustard" is from the old French "moustarde," which in turn came from the Latin "mustum," meaning "must." In this sense "must" refers to the fresh juice of grapes, or other fruit, with which the ground seeds of mustard were mixed to prepare a condiment.

Early colonists brought mustard to the Americas, and today it flourishes throughout the Western Hemisphere.

Mustard greens should be used when they are young and tender, either in a salad, or cooked as a potherb. To preserve the flavor and nutrients they should be cooked in a tightly covered pot, with only the water they are washed in left clinging to the leaves. A piece of salt pork or ham is an appetizing addition.

The condiment known as "mustard" is made by grinding the mustard seeds into a powder, and then mixing it with milk, water, or vinegar. It may also be used dry.

Still another type of Indian mustard seed is used to make oil; and there is the "medicinal" mustard seed, used in making mustard plasters.

Okra

"For corn the newcomers have a dogged aversion."

Thus did French settlers in Louisiana send back a hurry call to France for food, because the shipload of brides, known as the "Casket Girls," were staging a hunger strike in the roistering town of New Orleans in 1728. They were given that name because the all-alike trunks each girl brought with her were shaped like caskets. But corn they would not eat. Luckily for them, that glorious Louisiana concoction, gumbo, was waiting in the wings.

Okra, a basis for gumbo, came to this country with shiploads of Bantu slaves, and was first known as Kingumbo. Combined with *gumbo file*, a powder made from sassafras leaves, legendary gumbos were created, using chicken and oysters, veal and herbs, and one with seven greens and salt pork made especially for Holy Thursday.

Okra originated in the Abyssinian center which includes present-day Ethiopia. From there it was carried in prehistoric times to Arabia and North Africa.

It was taken to Egypt by the Moslems from the East when they conquered that country in the 7th century, and one of the earliest records of okra is a description of the plant by a Spanish Moor who visited Egypt in 1216. An interesting note is that he stated the young pods were eaten with meal, apparently in much the same manner as our own Southern fried okra dipped in corn meal is enjoyed today.

There are those who fail to appreciate okra, considering it "gooey" or pasty. Proper harvesting and cooking, however, eliminate most of this reason for objection. The pods should always be picked when very young. They grow very fast, and in hot weather develop from the pollinated flower in 3 to 5 days, after which they become tough. Cooking should be done very rapidly, and with care taken not to break the pods, as this preserves the flavor and texture.

Okra may be dried, and kept for later use in the same manner as fresh okra. Also, the ripe seeds of okra are sometimes roasted and ground to use as a substitute for coffee.

* * * * *

ROSELLE is a close relative of okra, and is used as a source of fiber for making cloth.

Onions

General Grant during the Civil War campaign in 1864 sent the following wire to the War Department: "I will not move my army without onions." The next day three trainloads of onions were started to the front. And Alexander the Great is said to have fed onions to his troops when in Egypt, to promote valor.

The onion has long been esteemed as one of man's most valuable foods, and history abounds with many such fascinating anecdotes of its merits.

The Bible states (*Numbers 11:5*) that the Israelites complained to Moses as they were being led out of Egypt: "We remember the fish which we did eat in Egypt freely; the cucumbers, and the melons, and the leeks and the onions and the garlick."

Herodotus, the Greek "Father of History" 5th century B.C. recorded that the workers engaged in the 20-year job of building the Cheops Pyramid consumed "onions, radishes and garlic costing 1,600 talents," (approximately 2 million dollars); and Hippocrates, Greek physician, wrote of the value of onions in 430 B.C. More than 1,200 years later, Chaucer, the English poet also wrote of them in his *Canterbury Tales*: "Wel loved he garleek, oynons, and eek lekes."

Onions have also figured as a symbol. When ancient Egyptians took an oath they placed their right hand on an onion, believing its formation of a sphere within a sphere represented eternity. And women of the Middle Ages frequently wore necklaces of onions to insure good health.

The onion derives its name from the Latin "unio." It is a bulb, and until recently was thought to be a member of the lily family. Experts in plant classification, however, now place it with the narcissus and amaryllis group.

Because of its widespread and ancient history, the place of origin of the onion is somewhat indistinct, although authorities favor Asia Minor. One exception is the Welsh onion (from the German "Walsch," meaning "foreign") which is native to China.

The onion was propagated throughout the globe, and today is a universal favorite.

The United States received onions by way of the West Indies, where they were taken by first Spanish voyagers.

There are many varieties of onion, including "Bermuda," "Creole," "White," "Yellow," and "Sweet Spanish." Most of them may be eaten either green or dried. They are unusually rich in valuable nutrients.

The Bermuda onion was once a big-money crop in the Rio Grande Valley. Then the market declined and a grower in Mission Valley, Texas, who shipped his bumper crop to market and got back a check for twenty-five cents, cashed the check, nailed the quarter to a mesquite tree, and switched to growing citrus.

Indiana grows profitable onion crops and in Walkerton, surrounded by peaty mucklands where onions thrive, a "mucklands crop fair" is held each year. Many of the onion farmers are Polish, Hungarian, and Lithuanians who seem to take naturally to "knee farming."

Before the development of the big, sweet Bermuda, onions were in demand in the West Indies, and a ship's captain, Isaac Bacon, living on Cape Cod, decided to go into the onion business. He grew excellent crops on his Cape farm and offered them for sale. Buyers wondered whether they might spoil before they reached the Indies. The captain assured them these were "tarnity onions ... they'll keep to all eternity." The captain would have had a rosy future on Madison Avenue.

Palmito, or Heart of the Palm

Palmito, or "heart of the palm," is just what the name implies, the terminal bud of certain palm trees, which, as it opens, forms the new fronds, or leaves. When this bud is cut out before it has developed, it is white and very tender; similar to the heart of celery in taste.

Palm trees are native to the tropics and subtropics, and this edible center of their leaves has been enjoyed by man in these parts from earliest times.

Palmito is not cultivated in the United States, but it is imported canned, and is a delicacy served either as a salad or heated as a vegetable.

Parsley

Curly-leaf parsley made gay and useful garlands for the ancient Greeks and Romans. Gay because the fresh green added a festive note to celebrations, and useful because the wearers could nibble on parsley sprigs in the belief that this would keep them from getting drunk.

Parsley has been found by modern experimenters to be a friendly plant, good for tomatoes when it grows near them and equally beneficial to roses.

Parsley is believed to have originated in southern Europe.

It is a member of the same family as celery, and the old Greek "selinon" mentioned by Homer about 850 B.C. in his *Odyssey*.

Theophrastus, Greek "Father of Botany," described several types of the plant in the 4th century B.C. Ancient Greeks used it extensively as a food flavoring. Early Romans also used parsley in their food, and are reported to have fed it to their chariot horses to make them swift.

Parsley was brought to the Americas in colonial days, and is at present popular throughout the world.

There are two general forms of parsley. The foliage type used for garnishing and flavoring, and the root type, which is cooked like other root crops.

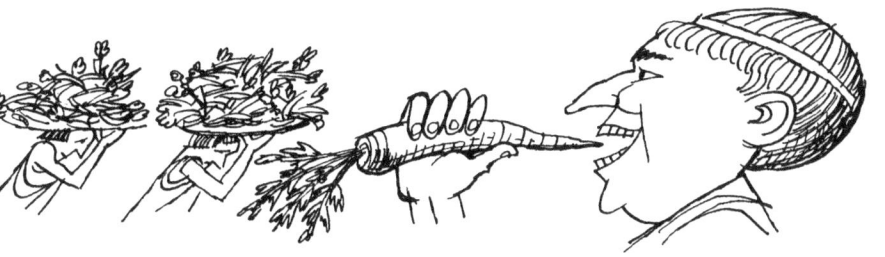

Parsnips

"Parsnips," states an old English cookbook, "are best left in the ground." But the parsnip, nevertheless, became a staple in Europe and was grown and cooked by the American Indians.

Parsnips are members of the carrot family, and are believed to have originated in the Mediterranean area and northeastward to the Caucasus. The Celts are credited with bringing them to Europe.

The Roman Emperor Tiberius, when he wasn't throwing slaves off the cliffs of Capri, was eating parsnips. He was so addicted to them that he had them imported from the Rhine country of Germany. The Romans believed that parsnips had medicinal value as well as food value. They did not cultivate them, however, but, rather, gathered them wild.

By mid-16th century parsnips were a common staple vegetable throughout Europe, much the same as the potato is today.

Early colonists brought them to this country, and at present they are cultivated rather extensively. They also grow wild. However, many of the wild varieties are poisonous. Wild parsnips, therefore, should not be gathered except by persons who are experts in distinguishing between the poisonous and the nonpoisonous kinds.

When properly cooked, parsnips have a sweet nutty flavor. They should always be steamed, and never boiled.

Potatoes

Perhaps Marie Antoinette never did say, "Let them eat cake," but when she wore potato flowers in her hair at the court of Louis XVI, she as much as said, "Let them eat potatoes."

The king had taken a fancy to that immigrant vegetable from Peru, the white potato, and decided that it ought to be a staple part of the diet of France. Not only did he persuade his queen to wear the blossoms, he also gave an all-potato banquet for such notables as Benjamin Franklin and Lavoisier. The court dutifully ate potatoes and wore potato flowers, but nothing happened. The French peasants regarded the potato as a dangerous relative of the mandrake, the deadly nightshade, and the lethal henbane.

Louis XVI, showing greater perspicacity than he did when the French Revolution was on his palace steps, decided to plant a potato field and place around it an armed guard during the day. He figured that anything so well guarded would appeal to the peasants as valuable. Cannily, he withdrew the guard after dark and sure enough, the fields were depleted night after night! Later, when soup kitchens were being set up all over Paris, a Frenchman named Parmentier had the soup made of potatoes, and thus gave his name to a classic dish.

The white potato that arrived in France in the 1770's had long been a valued plant in Peru and Bolivia, where Indians had developed it in a dozen different varieties and colors. These same Indians produced potatoes that were frost-resistant, a beetle-immune variety, and a type that was edible after freezing. Mexican and Peruvian markets still display potatoes in polka dots and stripes as well as some with purple skins and pale ivory flesh.

The Indian word "batata" used by settlers in North America actually applied to the sweet potato they found flourishing in native gardens. One of these sweet "batatas" was served to a man named John Pory in Virginia in 1621. It was roasted, he burned his tongue, and his ultimatum was: "I wouldn't give a farthing for a shipload."

Yet today the potato is probably the single most important vegetable in the world.

A favorite story is that Sir John Hawkins first brought the white potato to Ireland in 1565 while others insist that Sir Walter Raleigh took it there in 1585. At any rate, it became so important to the economy of that country that it acquired the name that has stuck to it ever since, "Irish potato." Irish immigrants brought it to this country in 1719, and the potato was first grown in Londonderry, New Hampshire.

Once introduced to the United States, the Irish potato quickly became a favorite. Some settlers, not knowing what to do with it, at first tried serving boiled potatoes with sugar and spices.

Since its pioneer days, the white potato has undergone the usual American transformation effected by our expert agriculturists. Idaho celebrates Spud Day in October, commemorating the coming to fame of its potato on the early Western railroads where it was served baked to bursting whiteness and drenched with butter. Up in the Aroostook County of Maine potatoes have become an equally well-known and profitable crop, with Long Island and California potatoes right up there in popularity. Yet only a little more than 100 years ago families pushing west in covered wagons paid $15 a bushel for potatoes in Central City, Colorado. San Francisco 49'ers paid a dollar apiece for potatoes, cooked, and an extra dollar for butter.

Pumpkins

One female pumpkin
An iron kettle
Salt and milk

Every bride marrying into the family of Putnams since colonial times had to learn to identify a tender female pumpkin and from it to make the family's traditional pumpkin bread. Just as the male squash blossom is considered best for eating (batter-fried), so the female pumpkin is less stringy, more delicate, and so is favored for pies, puddings, and soups.

That the pumpkin was everywhere when first settlers arrived in New England is attested by a ditty of the 1630's, which ran:

"We have pumpkins at morning and pumpkins at noon,
If it were not for pumpkins we would be undoon."

There was no cause, however, to be "undoon," for pumpkins had been growing for centuries in the Americas. In Circleville, Ohio, a pumpkin festival each October commemorates the earthworks upon which the town is built, where the Hopewell culture is believed to have grown pumpkins 1,000 years ago.

Pumpkins belong to the same family as squashes, and in fact botanically there is no accepted basis for distinguishing between them.

They are native to the Americas. Most varieties are declared to have originated in Mexico, Central and South America, but fragments of pumpkin stems and seeds have also been found in the ruins of the ancient cliff dwellings in southwestern United States.

The word "pumpkin" is derived from the old French "pompion," meaning eaten when "cooked by the sun," that is, ripe. The name is applied to the large round orange-colored fruits used in pies and Halloween lanterns, but many of the so-called "pumpkins" are in effect squashes.

Pumpkins usually grow on long, thick vines, but one variety, Cheyenne Bush Pumpkin, which was developed in recent years, grows on plants similar to the bush summer squash.

Pumpkins are nutritious, and very appetizing baked, mashed, scalloped with marshmallows and brown sugar, and in pies. And of course what would Halloween be without those pumpkin lanterns!

Radishes

Ancient Greeks valued radishes so highly they made small replicas of them in gold. They called them "raphanos," meaning "easily reared."

They were also a common food in Egypt long before the Pyramids were built, and Roman writers at the beginning of the Christian era described them.

China is said to be the country of origin of the radish, and from there it was introduced into middle Asia in prehistoric times. It derives its name from the Latin word "radix," which means "root."

Varieties of radishes range from the small, round, or long red and white ones, popular in Europe and the Americas, to the very large species, often as big as basketballs, grown in the Orient.

The mammoth type was apparently known in Europe long before the smaller ones. German writers of the 13th century described them, and one of their botanists later in 1544 reported radishes weighing 100 pounds!

The larger varieties are favored in the Orient, especially in China and Japan, where they are cooked, or pickled in much the same manner that we prepare cucumbers.

In Egypt and the Near East radishes are cultivated for their leafy tops, and India grows one variety for its edible pods.

Columbus and his followers brought radishes to the New World, and they were in the first gardens of our early colonists.

Radishes are tasty, nutritive, and easy to grow in a home garden.

Radish rosettes are gay for garnishing salads and cold dishes. Easy to prepare too. Just make thin cuts in the red skin from the top almost to the stem base. Drop into iced water, and the petals will soon curl back.

Rhubarb

Americans used to call it "pieplant" because the delicate pink-toned stalks make a delicious pie. But the rest of the rhubarb plant is a deceiver. The leaves are actually poisonous, and have been known to cause death, while the roots have been used medicinally in China and Asia since 2700 B.C. The root of the Chinese-type rhubarb, in fact, is still used in medicines today.

The word "rhubarb" comes from the French "rhubarbe," which is a contraction of the Late Latin term "rheubarbarum."

There are records of many species of rhubarb from various regions of Europe and Asia, but nothing specific is known about their relationships or origins. Our most popular varieties are believed native to the eastern Mediterranean and Asia Minor. Other edible species are found wild in China and in middle Asia.

The rhubarb we know in our markets today is different from its Oriental ancestor, and we do not use the roots at all, only the succulent pink stalks. It dates back to 17th-century Italy, from where it traveled to England in the late 1600's. It did not make its American debut until the time of the Revolution, when a Maine farmer introduced it to New England market gardeners. An American seed catalogue listed rhubarb in 1828. New England was ideal for rhubarb growing, since this is one plant which fancies a cool climate and must have a winter rest if it is to thrive. Interest in its cultivation is widespread today.

Rutabagas

The rutabaga is believed to be a hybrid, that is, a mixture of the turnip and some form of cabbage. It is not known exactly when this occurred, but the new species appeared in Europe in the late Middle Ages. Caspar Bauhin, Swiss botanist, first recorded the new vegetable in 1620.

It derives its name from the Swedish "rotabagge," and in many places is commonly called "Swedish turnip."

Rutabagas were grown in the royal gardens in England in 1664, although they were little known in the rest of that country at that time. They were apparently raised on the Continent at a much earlier date.

Rutabagas require a longer growing season than turnips, and a cooler climate. They are both prepared in much the same manner, but the rutabaga is considered more nutritious.

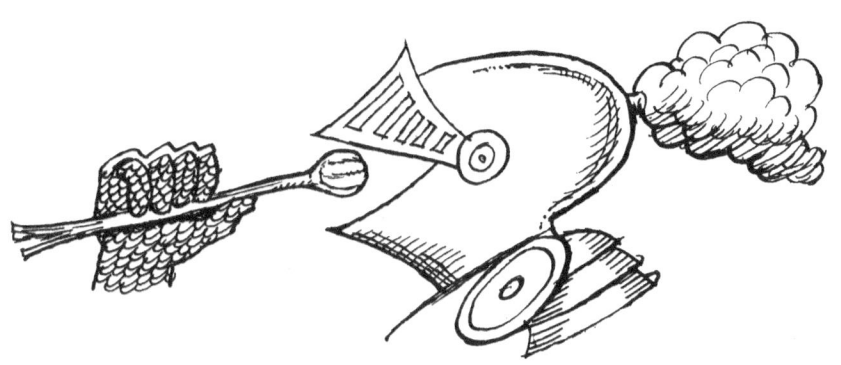

Shallots

The shallot belongs to one of the oldest groups of edible bulbs, and is related to onions, leeks, and garlic.

It is sometimes erroneously called "scallion," which is a shoot from the white onion that is pulled before the bulb has formed. Shallots, on the other hand, have a distinctive bulb made up of cloves similar to garlic, except that the individual cloves are not encircled by membrane.

The exact origin of shallots is undetermined, but they have long been the gourmet's delight. The French are credited with introducing them into Europe early in the Christian era. French knights returning from the Crusades (1095–1291) are said to have brought them from Syria.

They were introduced into the United States by early French colonists in Louisiana, and today, because of their delicate flavor and mild aromatic properties, are favored by those who prefer a lighter touch than garlic or onion as a seasoning.

Shallots are in the market during most of the year, either green, or cured in dry form.

Soybeans

The God Farmer, Emperor Sheng Nung, was the first to record the characteristics of the soybean. He did this in the year 2383 B.C. in his Chinese materia medica, *Pan Tsao Gong Mu*, making the soybean one of the world's oldest vegetables. The old Chinese name for it was "sou," from which were derived the various names "soi," "soy," and "soya."

So important is this "almost perfect food" that Dr. A. A. Horvath, who did extensive research in its nutritive values, believes that "the Chinese nation exists today because of the use of the soybean as food."

The soybean was introduced into North America by a ship's captain in 1804 and Commodore Perry is said to have brought some seeds back from Japan with him in 1854. However, people paid little attention to it until the rigors of World War II forced the United States into large-scale production. In about 35 years soybeans have mushroomed from an almost unknown forage crop into an annual 700 million bushels.

Thus, a vegetable nearly as old as civilization has become one of our newest and most exciting nutrition discoveries, and in addition has many commercial uses, such as in plastics, etc.

Spinach

A cartoon character named Popeye was responsible for a generation of American children eagerly devouring tons of spinach. Popeye ate spinach right out of a can, to give him strength, and invariably went into battle to win an impressive victory. In gratitude, the spinach growers of a Texas community erected a Popeye statue of heroic size.

The meteoric rise of spinach-eating coincided with a general American belief in the value of green vegetables. Spinach had been around for a long while, probably arriving in the United States with the early colonists. It was found on hotel menus in the early 1880's but usually served in a mashed, creamed form with an egg on top, and playing third or fourth best to turnips, squash, and potatoes.

Spinach is native to Iran (Persia) and nearby areas, but was not known in other parts of the world until about 2,000 years ago.

The oldest record we have of spinach is in Chinese, which says that it was introduced into that country from Nepal in A.D. 647.

From Syria and Arabia it reached North Africa, and the Moors took it from there to Spain around A.D. 1100. It was unknown, therefore, to the Greeks and Romans.

Our name "spinach" comes from the old French "espinache," which in turn was derived from Arabic and Persian names.

At present, spinach is probably the most important crop grown for greens in the United States.

Squash

Squashes are members of the large gourd family, to which pumpkins, watermelons, and cucumbers also belong. They are native to the Western Hemisphere, and were an important food for the Indians centuries before the white men arrived.

Although there are many names for the different varieties, our common word "squash" is derived from "askutasquash," used by the Indians of the North Atlantic coast, meaning "eaten raw and uncooked." We prefer to cook our squash today, but we still harvest "summer squash" while tender and unripe.

The innumerable types of squash may be divided into two general categories. The late-growing, or "winter squashes," with odd shapes and colors, thick shells, and a marked seed cavity; and "summer squashes," which are small, quick-growing, and are eaten before the rinds and seeds begin to harden. Pumpkins belong to this group, but many other late forms are mistakenly also called "pumpkins." Today, both the "winter" and "summer" varieties are available in the market almost all year.

Mexico and Central America are said by authorities to have been the center of origin of squashes, with some of the larger species, such as hubbard and marblehead, originating in the Andes Mountains of Peru and Argentina. Archaeological finds and records of the cultivation of squashes in all of these areas date back to 3000 B.C.

Early Spanish explorers carried the seeds north into what is now the United States, and also to Europe. Some of the newer forms, developed from these early types, have

been given later names, such as the "cocozelle" and "zucchini," bred in Italy.

Summer squashes, including the crookneck, straightneck, white bush scallop, cocozelle, and zucchini are delicate and tasty boiled, steamed, baked, or fried.

The winter, or hard-shell types, such as hubbard, Turk's-turban, and acorn, are usually cut in half and baked. They may be stuffed with various fillings, or served with brown sugar, butter, and spice.

All types of squashes are rich in nutrients, and they are an important staple food in many areas, especially throughout Central America and Mexico. The flowers of some of the varieties are edible and were once used in ancient Indian religious rituals. The mature seeds of others are dried and considered a delectable relish.

Sweet Potatoes

During the Revolutionary War a British officer came to discuss an exchange of prisoners with Francis Marion the famous "Swamp Fox" whose tactics so often baffled the Red Coats. He found Marion and his men deep in a South Carolina swamp, subsisting on sweet potatoes which they roasted on sticks over an open fire. Returning, he reported: "Such a people could not and ought not to be subdued."

Long before white men arrived in the Western Hemisphere, the Inca Indians of South America and the Mayas of Central America grew several varieties of sweet potatoes. Some were used for food, and others for colors in their paints. They called the plant "Cassiri."

Columbus ate roasted sweet potatoes upon his arrival in the West Indies. They were served to him by natives he described as "very gentle and courteous."

Early Spanish and Portuguese explorers took sweet potatoes to Europe, India, China, and Malaya.

The sweet potato, of which there are some 1,000 species, is one of the most complete foods known. It is an important staple nutrient in the warmer climates where it thrives, and is shipped from there to nearly every section of the world.

The flesh of the sweet potato varies from nearly white to almost orange in color. The darker varieties are often erroneously called "yams." The yam, however, is an entirely different plant, native to the tropics, and is rare in the United States.

The blossoms of the sweet potato plant look very much like those of its relative, the morning-glory. It blooms profusely in tropical climates, but rarely in our cooler United States.

Sweet potatoes are also used as a food for livestock, and their high starch content is valuable in many products.

Recently, the Russians and Australians have been experimenting with hardy new breeds, but in general the sweet potato is prized in the Americas where it is used as a vegetable, made into puddings, pies, and even soup and ice cream.

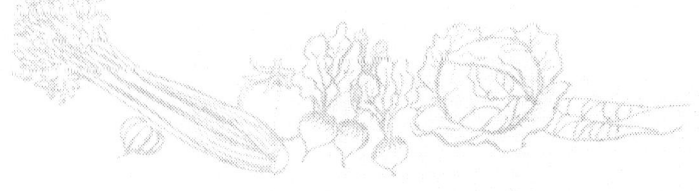

Swiss Chard

Swiss chard is a primitive, leafy species of the beet family. It has been known to man for hundreds, and some types for thousands, of years.

Theophrastus, Greek "Father of Botany" (37–287 B.C.), mentioned the green varieties, and Aristotle, Greek philosopher (384–322 B.C.), wrote of red chard. The Romans recorded it.

Swiss chard was introduced into Europe from Italy, and has been popular there for many centuries. Early settlers brought it to the New World.

The leaves may be cooked as greens, and the delicate stalks like asparagus. It is a rich source of vitamins.

Tomatoes

It took the United States Supreme Court to establish the tomato as a vegetable. In 1893 an importer argued that the tomato was a fruit, and because of that, not subject to duty. The Supreme Court held that the tomato was a vegetable because it was served in soup, with soup, or with the main course of the meal.

That is just one episode in the history of the long-suffering tomato. It was known for a time as the "love apple," probably because its Italian name, *pomo de mori* (Moorish apple) was perverted into *pomo d'amore.*

Just as the potato was thought to be poisonous because of its association with mandrake, so the tomato acquired a similar stigma because of its association with the family of the deadly nightshade.

The first tomatoes probably grew wild in the Andes. They are cultivated today throughout Peru, Bolivia, and Ecuador, where wild tomatoes also abound. Plant hunters have traced the tomato into Central America and Mexico, and they believe it was carried by man in much the same fashion as was corn. In North America, Indians were not growing tomatoes when white men arrived, possibly because the tomato is sensitive to extremes of climate.

Thomas Jefferson was growing tomatoes in 1781 and it is said that Philadelphia had its first crop in 1789, thanks to a French visitor from Santo Domingo, while Salem, Massachusetts, acquired tomato plants via an Italian painter in 1802.

Nevertheless, in the year 1830, a determined gentleman named Robert Gibbon Johnson stood on the steps of the courthouse in Salem, New Jersey, and ate a tomato before an awestruck crowd, to prove it really was good eating, not poison.

Now, of course, we drink most of our crop in tomato juice, and eat another large part in tomato soup. Back in 1897 a chemist, John Dorrance, took a $7.59 a week job with a small factory in Camden, New Jersey. One day he told his boss that he had worked out a formula for condensing and canning tomato soup. Thus was Campbell's soup born! Today, thousands of acres of tomatoes are grown solely for canning.

The next revolution came in the 1930's when people began *drinking* their tomatoes. And as America became more salad-conscious and more devoted to pizza and spaghetti, the tomato came ever closer to winning top popularity honors in the vegetable world.

Turnips and Turnip Greens

Ancient Greeks valued turnips so much they made small replicas of them in lead.

Turnips belong to the mustard family, and are believed to be native to Russia, Siberia, and Scandinavia. The plant grew so easily, however, and in so many places, that it was soon widely distributed.

The Romans and Greeks enjoyed turnips. Pliny, Roman naturalist in the 1st century A.D., described several kinds. They also spread to the Orient, where they continue to be a popular favorite.

In England, during the time of Henry VIII, turnip roots were boiled or baked, and the tops cooked as "greens." The young shoots were used in salads.

Jacques Cartier, the French navigator who discovered the St. Lawrence River in 1536, planted them in Canada in 1541. They were also carried south, and since colonial times have been one of the commonest and most popular garden vegetables in America.

Turnips are raised primarily for their roots, but the leaves are also very good when young and tender. Some kinds, in fact, are grown only for greens. And for centuries turnips have also been raised for stock feed.

Watercress

The ancient Persians were advised to feed watercress to their children if they wished to improve their growth, and the Greeks had a proverb which said that it was a "wit producing food."

Watercress is native to Asia Minor and the Mediterranean area.

Xenophon, ancient Greek general and historian (434–355 B.C.), and Xerxes, Persian king (480 B.C.), recommended watercress in the diet of their soldiers, since they observed that those who ate it kept in better health.

The Romans were fond of watercress, and used it in their salads.

Sixteenth-century English writers recommended it as a remedy for scurvy, and also as a beautifier for ladies.

The health-giving properties that the ancients attributed to watercress are apparently well founded in the light of modern scientific analysis, as it is now used in the treatment of many illnesses, principally tuberculosis.

The name "cress" is derived from the Middle English "cres" or "cress." Cresses, both land and water, stem from the mustard family, and are known as the "herbs of the mustard group."

Yams

The yam is native to the tropics, and is believed to be African in origin.

Yams include some 250 species of climbing vines, some of which have soft fleshy roots which are edible. These are the ones which are relished as a tasty food.

The darker, moist varieties of sweet potatoes are often erroneously called yams in the United States.

Yams are a favorite food in the tropics. They are prepared in the same ways as is the sweet potato.

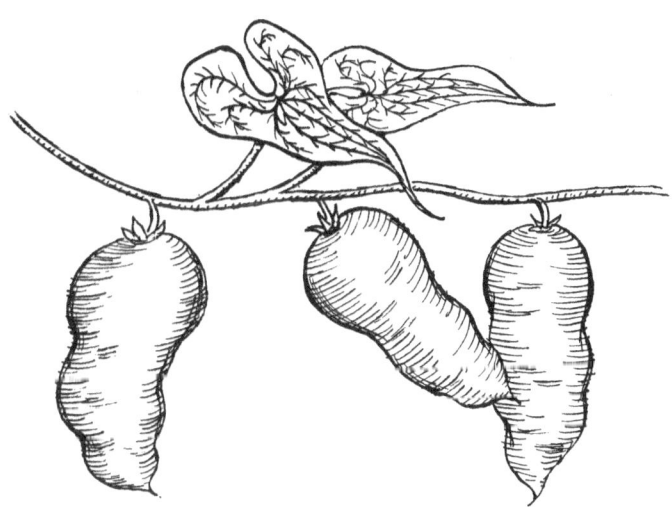

Bibliography

Sturdevant's *Notes on Edible Plants*
Botanical-Geographic Principles of Selection, by N. I. Vavilov
The Origin of Indian Corn and Its Relatives, by P. C. Mangelsdorf and R. G. Reeves
Our Vegetable Travelers, by Victor Boswell (principal horticulturist U. S. Department of Agriculture)
Staircase Farms of the Ancients, by O. F. Cook
New Plant Immigrants, by Donald Fairchild
Bulletins from the "United Fresh Fruit and Vegetable Assn.," Washington, D.C.
Mushroom Growing Today, by Frederick Atkins
Hunting Useful Plants in the Caribbean, by David Fairchild
Mary Margaret McBride's Harvest of American Cooking, G. P. Putnam's Sons, New York, 1957
Southern California Country by Carey McWilliams, Duell, Sloan and Pearce, New York, N. Y. 1956
Along the Maine Coast, by W. N. Wilson, Whittlesey House, McGraw-Hill, New York, 1947
Indiana, the American Guide Series, Oxford University Press, 1941
The Agricultural Regions of the United States, by Ladd Haystead and Gilbert C. Fite, University of Oklahoma Press, Norton, Okla., 1955
Cape Cod's Way by Scott Corbett, Thomas Y. Crowell, New York, 1955
The World in Your Garden, National Geographic Society, Washington, D.C., 1957
Plant Drugs That Changed the World by Norman Taylor, Dodd, Mead & Co., New York, 1965
History and Development of the Citrus Industry, by Herbert John Webber in The Citrus Industry, Volume I, available from University of California press.
The *Encyclopaedia Britannica*
The *National Geographic Magazine*
The Art of American Indian Cooking by Yeffe Kimball and Jean Anderson, Doubleday & Co., Garden City, N.Y. 1965

INDEX

Ababai, 62
Adam's Apple, 46
Afghanistan, 94
Africa, 24, 25, 80, 103, 155
Ahuacatl, 24
Akbar, Emperor, 53
Alcott, Louisa May 18
Alcinous, Garden of, 67
Alexander the Great, 21, 25, 129
Algonquin Indians, 114
Alligator Pear, 23
Amadas, Philip, 47
American Indians, 71, 101, 113, 114, 121, 124, 133, 151
Anana, 70
Andes Mts., 100
Aphrodite, 17
Apple Blossom Festival, 20
Apple of Discord, 17
Apples, 17, 59
Apple Plum, 71
Appleseed, Johnny, 19
Apple Slump, 18
Apricots, 21
Arabs, 25, 40, 49, 51, 80
Aristotle, 93, 150
Arizona, 40, 46, 57, 120
Armenia, 55
Aroostook County, Me., 135
Artichokes, 84
Asia, 17, 28, 34, 38, 44, 49, 53, 73, 74, 79
Asian Pear, 67
Askuta Squash, 146
Asparagus, 86
Astor House, 43
Athena, 17
Augustus, Emperor, 86
Australia, 149
Avocados, 23
Aztec Indians, 23

Bacchus, 42
Bacon, Isaac, 131
Bahrein Island, 40
Baker, Capt. Lorenzo, 26
Baldwin Apple, 19
Bananas, 25
Banana Melon, 55
Barbados, 46
Barlow, Arthur, 47
Batata, 135

Bauhin, Caspar, 142
Beans, 103, 113
Beets, 87
Belgian Endive, 106
Bemuth, Charles, 105
Bermuda Onion, 130
Bing Cherries, 33
Bio-Dynamic Farming, 92
Bitter Berry, 36
Blackberries, 28
Black-Eyed Peas, 103
Black Sea, 48
Blackstone Pudding, 18
Black Currents, 39
Blastophaga, 43
Blue Berries, 30
Boston Lettuce, 120
Brazil, 50, 53, 61, 70, 80, 122
Bread Fruit, 48
Bronze Age, 48, 111
Boston, Mass., 18
Bo Tree, 42
Boysenberry, 29
Britons, 86, 89, 118
Broccoli, 89
Brussels, Belgium, 91
Brussels Sprouts, 91, 116
Buddhism, 42, 53
Bull, Ephriam, 48
Burbank, Luther, 71
Burdick House, Mich., 96
Burma, 60, 126
Butter Beans, 122

Cabbage, 92, 114
California, State of, 19, 32, 40, 48, 50, 65, 71, 120
Cambridge, Mass., 77
Camden, New Jersey, 152
Canary Islands, 25, 50, 52, 60
Canada, 72, 153
Cantaloupe, 55
Cantalupo Castle, 55
Canterbury Tales, 117
Cape Cod, Mass., 26, 100, 131
Caprific, 43
Cardoon, 84
Caribbean Area, 26, 70
Carrots, 94
Carthaginian Medals, 73
Cartier, Jacques, 93, 153

Casaba, 55
Cascades, 20
Casket Girls, 127
Caspian Sea, 17, 48
Cassava Plant, 122
Cassiri, 148
Cato, 17
Cauliflower, 95
Caulis, 115
Caves, Prehistoric, 32
Cayenne Pepper, 108
Celeriac, 97
Celery, 96
Celts, 93, 99, 114, 133
Central America, 24
Central City, Colorado, 136
Cepa, 118
Champignon, 125
Chapman, John, 19
Charlemagne, Emperor, 104
Charles the II, 36
Chaucer, Geoffrey, 117
Cherokee Indians, 30
Cherries, 32
Cheyenne Bush Pumpkins, 138
Chicory, 98
Chile, S.A., 78
Chili Pepper, 108
China, 21, 25, 32, 49, 59, 60, 64, 68, 72, 79, 80, 94, 96, 104, 109, 126, 140
Chinese Gardners, 97
Chives, 117
Cider, Apple, 20
Circleville, Ohio, 138
Citrus Aurantium, 61
Claudia, Queen, 72
Climax Plum, 71
Coachella Valley, Calif., 40
Coconuts, 34, 59
Coconut Oil, 34
Cocozelle, 147
Coffee Substitute, 40
Colewarts, 99, 115
Coles, 115
Collards, 99
Columbus, Christopher, 34, 50, 52, 54, 60, 70, 104, 107, 119, 140
Concord Grape, 48
Concord, Mass., 18
Connecticut, State of, 18
Conquistadores, 70, 101
Corn, 100, 114

157

Corn Palace, S.D., 102
Corinth Grape, 39
Cotton, 59
Coville, Dr. Frederick V., 31
Cow Peas, 103
Crab Apple, 20
Cranberries, 36
Creek Indians, 101
Crispo, Aniello, 120
Croatan, 47
Crook Neck Squash, 147
Cross-berry, 44
Crusades, The, 49, 51
Cucumbers, 80, 104
Cucumis, 54, 104
Cumberland Experimental, 65
Currents, 38
Cydon, Crete, 74

Damascus, Syria, 72
Damson Plum, 72
Dates, 40
Dehydration, 87
de Berlanga, Friar Tomás, 26
Delaware Pink, 48
Delmonicos Rest., 23
de Maintenon, Madam, 111
Dennis, Cape Cod, 37
Dew Berries, 28
Dorrance, John, 152
Downing, Lewis, 124
Drupes, 22
Dunclas County, Ont., 19

Eggplant, 104, 105
Egypt, 42, 52, 64, 73, 80, 84, 92, 127, 129, 139
Elberta Peach, 64
Endive, 106
England, 44, 47, 67, 89, 143
Ericsson, Leif, 100
Escarole, 106
Eskimos, 30
Espinache, 145
Ethiopia, 127
Europe, 28, 38, 42, 44, 77
European Pear, 67

Fannie Heath Raspberry, 76
Field Beans, 113
Figs, 42
Florida, 23, 45, 53, 60
Fossils, 42
France, 42, 67, 78, 84, 85, 88, 104, 111, 143
Francis, 1st, 72
Franklin, Benjamin, 134
Freezing, 17
French Endive, 106
Frezier, Capt., 78
Fruta Bomba, 62
Fox Grapes, 30
Fungus, Gooseberry, 44

Gage, Sir Wm., 72
Galen, 54
Garden of Eden, 25, 42
Garlic, 104, 109, 143
Germany, 67, 88, 111, 133, 139
Gold Plum, 71
Gold Rush, 19
Golden Delicious, 20
Golden Jubilee, 65
Gooseberry, 38, 44
Grand Banks, 26
Grant, General U. S., 129
Grapefruit, 45
Grapes, 47, 59
Great Lakes, 48, 65
Greece, 17, 21, 39, 54, 56, 65, 67, 74, 84, 92, 94, 96, 99, 111, 114, 118, 132, 153
Green Beans, 113
Green Gage Plum, 72
Green Peas, 111
Green Olives, 58
Grenadine, 73
Griffiths, Billie, 49
Grits, 114
Ground Cherries, 30
Guadeloupe Island, 70
Guarani Indians, 70
Guatemala, C.A., 24, 100, 121, 123
Guatama, Siddartha, 42
Gumbo, 127

Half Moon Bay, Calif., 85
Hall, Henry, 37
Hawaiian Islands, 24, 70
Hawkins, Sir John, 135
Heart of Palm, 131
He-i, 62
Helen of Troy, 17
Henry, VIII, King, 153
Hera, 17

Herbert, Sir Thomas, 51
Herodotus, 119
Himalayas, 71
Hindustani, 105
Hippocrates, 129
Hispaniola, 26, 43, 50, 52, 60
Holy Land, 58
Homer, 67, 73, 96, 132
Hominy, 114
Honeydew, 55
Hopewell Indians, 138
Horvath, Dr. A. A., 144
Hovey, Chales, 77
Huckleberry, 30
Hughes, Richard, 38
Hungary, 108, 111
Hush Money, 36
Hybridize, 91
Hyssop, 92

Iceberg Lettuce, 120
Idaho Potato, 135
Incas, 24, 100, 102, 148
India, 25, 65, 103, 105, 109, 119, 126, 147
Indian, American, 24, 36
Indian Mustard, 104
Iita, 62
Imperial Valley, Calif., 55
Iran, 73, 74, 145
Ireland, 135
Irish Potatoes, 135
Italy, 21, 67, 84, 118

Jamaica Island, 26, 103
Jamestown, Va., 43, 65
Japanese, 68, 72, 79, 140
Japanese Carrot, 94
Japanese Flowering Cherries, 33
Japanese Flowering Quince, 74
Jefferson, Thomas, 152
Jemez Indians, 101
Jerusalem Artichokes, 85
Johnson, Robert Gibbon, 152

Kaki, 68
Kalamazoo, Mich., 96
Kale, 115
Kidney Beans, 113

Kingumbo, 127
Kohlrabi, 91, 116
Kruisbes, 44
Kuyung, 59

Lactuca, 119
Lavoisier, Antoine, 134
Lebanon, Ore, 77
Leeks, 104, 117, 143
Lemons, 49
Lentils, 118
Lettuce, 119
Licking Creek, Ohio, 19
Lima Beans, 113, 121
Lima, Peru, 121
Limes, 51
Lime Juice, 24, 36
Livingstone, Dr. David, 80
Loganberry, 29
Logan, Joseph K. H., 29
Londonderry, New Hampshire, 135
Long Island, N.Y., 135
Louis XIV, King, 111
Louisiana, 127, 143, 143
Love Apple, 151
Luelling, Henderson, 32

Mad Apple, 105
Magness, John R., 66
Magnus, Albertus, 60
Maize, 100, 114
Malay Archipelago, 25, 34, 46
Mandarin Orange, 79
Mandioca, 122
Mangoes, 53
Maraschino Cherries, 33
Marie Antoinette, 134
Marion, Francis, 148
Mark Twain, 95
Mandrake, 151
Martha's Vineyard, 100
Massachusetts Bay Colony, 38
Massachusetts, State of, 19, 48, 71, 77
Mayas, 24, 62, 100, 123, 148
Mayflower, The, 32
MacIntosh, John, 19
Meat Substitute, 24
Mediterranean Area, 17, 42, 43, 79, 84, 86, 87, 99, 109, 115, 117, 133
Melons, 54, 104

Melopepo, 54
Mexico, 23, 24, 52, 70, 102, 138, 146, 151
Miami, Fla., 24
Michigan, State of, 32, 68
Milikane, 62
Mission Fig, 43
Mission Valley, Texas, 130
Mistress Pearce, 43
Mitchell, South Dak., 101
Mobile, Ala., 26
Mohammed, 40
Mongolians, 49
Montezuma, Emp., 23
Montreal Melon, 55
Moors, 80
Morgan, B. J., 32
Moslem, 40
Moses, 55
Mount Ida, Greece, 75
Mozambique, 51
Mucklands Crop Fair, 130
Musa Paradisiaca, 25
Mushrooms, 123
Muskmelon, 54, 80
Mustard, Indian, 104, 126

Naples, Italy, 84
National Geographic Mag., 66
National Orange Show, 61
Navajos Indians, 101
Nectarines, 56
Nepal, 42, 145
Nero, Emperor, 117
New England, 26, 32, 37, 38
New Jersey, 31, 65
New Mission, Texas, 45
New Orleans, La., 26, 79, 98
New World, 18, 24, 26, 111
Niagara White, 48
Noah, 48, 57
Norman Conquest, 111
Norse Men, 47
North America, 28, 36, 38, 44
North Carolina, 20, 47

Odyssey, The, 67, 72, 132
Ohio River, 19
Okanogan Area, 20

Okra, 127
Olea Europaea, 57
Olives, 57, 59
Onions, 104, 118
Ontario, Canada, 19
Orange County, Calif., 97
Oranges, 59, 79
Oregon, State of, 19, 32

Pacific Islands, 25, 80
Palestine, 51
Palmito, 131
Papain, 63
Papayas, 62
Paprika, 108
Paraguay, 70
Parsley, 96, 132
Parsnips, 133
Parson Weems, 32
Paw Paw, 48, 62
Peaches, 59, 64
Pears, 59, 67
Peas, English, 103, 111
Penn, William, 64
Peppers, 107
Peppermint, 92
Perrine, Henry, 24
Perry, Comm. John, 69, 144
Persia, 40, 49, 51, 54, 55, 73, 74, 94, 104, 154
Persian Apples, 65
Persian Melon, 55
Persimmons, 68
Peru, 24, 107, 134, 146, 151
Phoenicians, 73, 86
Pie Plant, 141
Pimiento, 108
Piña, 70
Pineapples, 70
Pink Lemonade, 49
Pinto Beans, 113
Plantin, 25
Plata, 24
Pliny, 17, 21, 33, 67, 75, 89, 95, 153
Pliny, The Elder, 54, 72, 74, 77
Plums, 71
Plymouth Colony, 100
Pod Peas, 112
Polynesia, 24
Pomegranates, 59, 73
Pompeii, 74
Pompion, 138
Popcorn, 102
Portuguese, 25, 50, 53, 62
Pory, John, 135
Potatoes, 134
Protein, 24

159

Providence, Rhode Island, 18
Prunes, 71
Prunes Armeniaca, 21
Pummelo, 46
Pumpkins, 137
Putchamis, 68

Quince, 74

Rabelais, Francois, 119
Radishes, 139
Raleigh, Sir Walter, 47, 135
Raphanos, 139
Raspberry, 28, 75
Reine Claude Plum, 72
Rhubarb, 141
Rio Grande Valley, 130
Roanoke, Va., 47
Rockahominy, 114
Rodin, Jean, 78
Romaine, 120
Romans, 25, 42, 56, 57, 67, 84, 86, 109, 111, 114, 117, 120, 132, 153, 154
Romeo, Mich, 65
Roselle, 128
Rosemary, 92
Royal Anne Cherries, 33
Rumph, Samuel, 64
Russia, 80, 149
Rutabagas, 142

Sage, 92
Salem, Mass., 152
Salem, New Jersey, 152
Salinas, Calif., 120
Sand Pear, 67
San Francisco, Calif., 65, 85
Sanskrit, 25, 65, 73, 80, 103, 105
Santa Claus Melon, 55
Santa Rosa, Calif., 71
Savannah, Georgia, 59
Scandinavia, 32, 37
Scuppernog, 47
Scurvy, 36, 51, 154
Selinon, 96, 132
Shaddock, Capt., 46
Shallots, 143
Sheng Nung, Emperor, 144
Smallage, 96
Smith, Captain John, 21, 43, 68

Smyrna Fig, 43
Snap Beans, 113
Song of Solomon, 17
South America, 24, 29, 70, 77, 138
Soybeans, 144
Spain, 24, 46, 70, 78, 84, 95, 105, 130, 149
Spanish Missions, 22, 48, 50, 52, 57, 60, 65, 67
Spinach, 145
Spud Day, 135
Squash, 80, 146
Starch, 70
St.Augustine, Fla. 43, 50, 52, 60
St. David, Order of, 117
Stone Age, 17, 32, 42
Straw Berries, 77
String Beans, 113
Succotash, 121
Sugar Beets, 87
Summer Squash 146
Sun-Drying, 17
Supreme Court, U.S., 151
Swamp Angels, 97
Swedish Turnip, 142
Sweet Potatoes, 148
Sweet Spanish Onion, 130
Swiss Chard, 87, 150
Switzerland, 72, 111
Syria, 57, 72, 95, 145

Tangerines, 79
Tangier, Morocco,79
Tannin,68
Tapioca, 122
Ta Yu, Emperor, 59
Taylor, James, 97
Telegraph, The, 26
Tenochtitlan, 23
Texas, 45
Theophrastus, 17, 32, 67, 132, 150
Thistle, 84
Thyme, 92
Tibbetts, Eliza, 60
Tiberius, Emperor 133
Toltecs, 24
Tomatoes, 151
Tomato Juice, 152
Turnips, 153
Tortillas, 24
Traverse City, Mich., 32
Tree Melon, 62
Trinity Vine, 48

Trojan War, 17
Tule-Rooters, 97
Tumbling, 35
Tupian Indians, 122

United Fruit Company, 26
U.S. Dept. of Agriculture, 31

Valley of the Indus, 25
Varro, Marcus Terentius, 33
Vinland, 47
Virgin, 65, 77
Virginia, 21
Vitamins, 39, 50, 51, 55, 95

Wales, 117
Walkerton, Indiana, 130
Washington, George, 32, 103
Washington Market, 120
Washington, state of, 20
Watercress, 154
Water Melons, 80
Wax Beans, 113
Welch Onion, 130
Wellfleet, Cape Cod, 26
Wenatchee Area, 20
West Indies, 46, 52, 63, 80
White Antibes Winter, 55
White Bush Scallop, 147
White, Elizabeth C, 31
White Onion, 130
Whites Bog, N.J., 31
Wickson Plum, 71
Wilmington, Mass., 19
Winesap, 19
World War II, 23

Xenophon, 154
Xerxes, King, 154

Yakima Area, 20
Yams, 149, 155
Yellow Newton, 20
Yellow Onion, 130
Young Berry, 28

Zucchini, 147
Zuni Indians, 101

www.ingramcontent.com/pod-product-compliance
Lightning Source LLC
LaVergne TN
LVHW040740250326
834688LV00031B/374